I0132254

Nuclear Ethics in the Twenty-First Century

Nuclear Ethics in the Twenty-First Century

Survival, Order, and Justice

Thomas E. Doyle II

ROWMAN &
LITTLEFIELD
INTERNATIONAL

London • New York

Published by Rowman & Littlefield International Ltd.
6 Tinworth Street, London, SE11 5AL, UK
www.rowmaninternational.com

Rowman & Littlefield International Ltd. is an affiliate of Rowman & Littlefield

4501 Forbes Boulevard, Suite 200, Lanham, Maryland 20706, USA
With additional offices in Boulder, New York, Toronto (Canada), and Plymouth (UK)
www.rowman.com

Copyright © 2020 by Thomas E. Doyle II

All rights reserved. No part of this book may be reproduced in any form or by any
electronic or mechanical means, including information storage and retrieval systems,
without written permission from the publisher, except by a reviewer who may quote
passages in a review.

British Library Cataloguing in Publication Data

A catalogue record for this book is available from the British Library

ISBN: HB 978-1-4422-7660-4

Library of Congress Cataloging-in-Publication Data
Names: Doyle, Thomas E., II, author.
Title: Nuclear ethics in the twenty-first century : survival, order, and justice /
   Thomas E. Doyle, II.
Description: London ; New York : Rowman & Littlefield Publishing Group, 2020. |
   Includes bibliographical references and index. | Summary: "Using a constructivist
   approach, the book addresses international security studies' concerns about the
   relevance of moral reasoning to strategic and political thinking"—Provided by
   publisher.
Identifiers: LCCN 2019049332 (print) | LCCN 2019049333 (ebook) |
   ISBN 9781442276604 (cloth) | ISBN 9781442276611 (epub)
   ISBN 9781538164136 (pbk)
Subjects: LCSH: Nuclear arms control—Moral and ethical aspects. | Nuclear weapons—
   Moral and ethical aspects. | Nuclear nonproliferation—Moral and ethical aspects. |
   Nuclear disarmament—Moral and ethical aspects. | Security, International—Moral
   and ethical aspects.
Classification: LCC JZ5665 .D68 2020 (print) | LCC JZ5665 (ebook) |
   DDC 172/.422—dc23
LC record available at https://lccn.loc.gov/2019049332
LC ebook record available at https://lccn.loc.gov/2019049333

Contents

Abbreviations

CFC	Chlorofluorocarbon emissions
CINCPAC	Commander-in-Chief, Pacific Command
CTBT	The 1996 Comprehensive Nuclear-Test-Ban Treaty
EU	The European Union
FMCT	Fissile Materials Control Treaty
HINW	Humanitarian Imperative to Abolish Nuclear Weapons
IAEA	International Atomic Energy Agency
ICAN	International Campaign to Abolish Nuclear Weapons
ICJ	International Court of Justice
ICRC	International Committee of the Red Cross
IFRC	International Federation of Red Cross and Red Crescent Societies
INF	Intermediate Nuclear Forces Treaty of 1987
IR	International Relations
JCPOA	The Joint Comprehensive Plan of Action. Also known as the Iran Nuclear Deal.
LP	John Rawls's *Law of Peoples* (1999)
MAD	Mutually assured destruction
N5	The five *de jure* nuclear-weapon states in the NPT: the United States, Russia, Great Britain, France, and China. Sometimes referred to as the NPT NWS.
NAC	New Agenda Coalition
NATO	North Atlantic Treaty Organization
NNWS	Non-nuclear-weapon states
NPDI	Non-Proliferation and Disarmament Initiative
NPR	Nuclear Posture Review
NPT	The 1968 Nuclear Nonproliferation Treaty

NPT NWS	The five *de jure* nuclear-weapon states in the NPT: the United States, Russia, Great Britain, France, and China. Sometimes referred to as the N5
NSNW	Non-strategic nuclear weapons
NWS	Nuclear-weapon states
OSCE	The Organization for Security and Cooperation in Europe
PL	John Rawls's *Political Liberalism* (1996)
PLO	The Palestine Liberation Organization
PMSI	The Principle of the Morality of Social Institutions
POTUS	The President of the United States
PP	Immanuel Kant's essay *On Perpetual Peace* (1795)
PrepCom	Preparatory Committee Meetings of the States-Parties to the Nuclear Nonproliferation Treaty
RAND	Think tank associated with the U.S. Air Force. Abbreviation for "**R**esearch **AN**d **D**evelopment"
RevCon	Review Conference of the States-Parties to the Nuclear Non-proliferation Treaty
TJ	John Rawls's *A Theory of Justice* (1971)
TPNW	The 2017 Treaty on the Prohibition of Nuclear Weapons
UN	The United Nations
UNSC	United Nations Security Council
U.S.	The United States
USSR	Union of Soviet Socialist Republics, also known as the Soviet Union
WIP	The Wrongful Intentions Principle
WMD	Weapons of mass destruction
WWII	World War II (1939–1945)

Preface and Acknowledgments

This book continues an ambitious effort that I launched as a PhD student in 2008 to rethink nuclear ethics in a post–Cold War era, whose nuclear disarmament hopes in the mid-1990s had faded in the wake of the September 11, 2001, terrorist attacks, the renewed fear of new nuclear proliferation, and the deepening mistrust between the world's nuclear-armed states. My two mentors at the University of California Irvine, Etel Solingen and Patrick Morgan, noticed my interdisciplinary interest in ethical theory and the subject of nuclear weapons policy in international security studies. Under their guidance, I learned that a substantial nuclear ethics literature had arisen during the Cold War era, although it had become dormant in the post–Cold War era.[1] It focused mostly on the morality of nuclear war and deterrence (see chapter 2 in this volume). For her part, Etel's research combined international political economy and nuclear proliferation studies in remarkably fruitful ways.[2] She encouraged me to see if there were new applications of nuclear ethics to subjects beyond nuclear war and deterrence. Her advice inspired me to concentrate on the ethical dilemmas that attended nuclear proliferation and the corresponding evolution of nonproliferation policy in support of the Nuclear Nonproliferation Treaty (NPT) regime. Additionally, I became interested in the ethical dilemmas of nuclear abolition.[3] I owe a deep debt of gratitude to Etel for suggesting this line of research and for her continual support of me and my work.

For his part, Pat Morgan encouraged me to not lose sight of the relevance of nuclear deterrence to nuclear ethics. Pat's expertise on deterrence theory is well known,[4] and his advice led me to think about the ethics of nuclear deterrence in diverse contexts. One context is that experienced by small power adversaries of the United States, such as Iran. If one assumes (as do many commentators) that nuclear deterrence is morally justified for the United

States, then it seemed fair to ask, On what grounds, if any, is nuclear deterrence not morally appropriate for U.S. adversaries, such as Iran? I argued that, while the NPT forbids any non-nuclear-weapon state party from acquiring nuclear weapons for any purpose, the same moral reasons that justify U.S. nuclear deterrence also justify Iranian nuclear deterrence if Tehran faces an existential threat from a regional or global nuclear-armed adversary.[5]

Another context that Pat's advice suggested to me is the interest expressed by some commentators in deterring nuclear terrorism by nuclear threats, which I argued is practically and morally problematic.[6] Specifically, I argued that holding terrorist groups' assets hostage to nuclear threats is practically impossible. In the case of al Qaeda, it would seem that the only fixed asset a nuclear-armed state might target would be Mecca. However, I argued that threatening Mecca with nuclear strikes to deter al Qaeda's nuclear terrorist ambitions is altogether beyond the pale. Beyond the question of deterring nuclear terrorism, I was also motivated to think more on the question of nuclear defense and deterrence commitments by liberal democratic states. I was influenced by Daniel Deudney's discussion of nuclear despotism, which suggested that liberal democratic commitments to nuclear defense and deterrence compromise not only essential democratic practices but also liberal democratic values and identities.[7] In this volume's fourth chapter, this theme is developed more fully. The final and broader context is that which this book addresses, in which the dominant security frameworks—that is, national security, alliance security, collective security—that continue to organize state interactions and international security studies itself are increasingly morally problematic. Accordingly, I want to thank Pat for helping me to stay closely tuned to the new developments in nuclear deterrence theory and policy study, as well as his indispensable advice in several other matters regarding advancement in the academy.

My effort over the past decade to rethink nuclear ethics owes a great debt of gratitude to the Cold War–era commentators—many of whom are cited in these pages—who developed the first lines of thinking about the morality of nuclear war and deterrence. The conventional obliteration bombing campaigns undertaken by the Allies in World War II established new thresholds for indiscriminate killing in modern warfare, and as a result pronounced disagreement among commentators arose over the moral appropriateness of the atomic bombings of Hiroshima and Nagasaki. Additionally, these Cold War–era commentators charted new paths in thinking about the moral permissibility or impermissibility of threatening with the use of nuclear weapons for the purpose of war prevention. Indeed, the moral paradox that the prevention of nuclear warfare might require the resort to credible threats of nuclear reprisal suggested to many, but not to all, commentators the impossibility of unqualified moral judgments on these policies. These debates are recounted in some detail in this book's second and third chapters, and one purpose of this recounting is to identify lines of security and ethical analysis that were not fully developed and might also provide useful grounds for new thinking in nuclear ethics. On the basis of this recounting, it is the broader purpose of this book to propose one line of morally informed security theory that suggests a new way of thinking nuclear ethics: common security.

In late 2015, I was approached by Marie-Claire Antoine, then working for Rowman & Littlefield Publishers, to gauge my interest in writing a monograph on nuclear ethics. I accepted their invitation and began the project in earnest in 2016 on the assumption that the Obama administration's commitment to nuclear disarmament would make incremental progress despite some frustrating setbacks, such as its failure to advance the cause of the Comprehensive Nuclear-Test-Ban Treaty in the U.S. Senate or its decision to proceed with nuclear modernization. Little did I anticipate the election of Donald J. Trump to the U.S. presidency in November of that year. Furthermore, I did not anticipate the directions in foreign and nuclear policy that the Trump administration took in 2017 and 2018. As I continued to reflect on the rapidly changing contexts of international security during this period of time, I came to believe that I could not proceed with the book according to my original designs. I contacted Marie-Claire and asked if I might revise my original book proposal and, if acceptable, receive an extension for the manuscript deadline. I received an affirmative answer to both requests, for which I am very grateful. I thank Rowman & Littlefield for their kind invitation to write this book. I thank Marie-Claire for her work on my behalf and thank Rebecca Anastasi, Dhara Snowden, and Natalie Bolderston for taking over this project when Marie-Claire's tenure at Rowman & Littlefield ended.

I also want to thank the following scholars and experts for reading one or more chapters of the manuscript in preparation for publication: Laura Considine, Antonio Franceschet, Harry Gould, Eric Heinze, Wade Huntley, Ulrich Kuhn, Tony Lang, Jennifer Mitzen, Cian O'Driscoll, Ionut Popescu, Nick Ritchie, Brad Roberts, Tom Sauer, Brent Steele, and Milla Valla. I also want to thank the anonymous peer reviewers at Rowman & Littlefield for their suggestions on the original book proposal, which I was able to apply even in the revised book proposal and finished monograph. Apologies in advance to anyone not mentioned here, who helped me with the manuscript directly.

Additionally, I want to thank those with whom I have had productive e-mail correspondence or face-to-face discussions (mostly in the hallways or conference rooms at various university settings or International Studies Association conferences), which have influenced my thinking on nuclear ethics but which did not offer direct input on this book. These scholars and experts include Daniel Brunstetter, Amy Eckert, Lewis Griffith, Marianne Hanson, Anne Harrington, Don Inbody, Aaron James, Nate Jones, Haider

Khan, Steven P. Lee, James Mark Mattox, Kevin Olson, James Pattison, Ben-oit Pelopidias, Maria Rost Rublee, Scott Sagan, Martin Schwab, Nina Tan-nenwald, Nico Taylor, Ramesh Thakur, Wilfred Wan, Nicholas J. Wheeler, and Robert E. Williams. Of course, I take full responsibility for any errors that remain in the book.

Finally, I want to thank my wife, Chrislea, and my children, Ely and Holly, for their continued and full support throughout the process of writing this book. I spent many a weekend devoted to various aspects of this book that could have been spent with them hiking or otherwise engaging the beauty of the central Texas Hill Country or the sights and sounds of Austin or San Antonio. I dedicate this book to them with the aspiration that we will live to see the dawn of an age free from the threat of nuclear war or accident.

NOTES

1. Doyle II (2010a).
2. See, for example, Solingen (2007).
3. Doyle II (2009, 2010a, 2015b, 2015c).
4. See, for example, Morgan (2003, 2009).
5. Doyle II (2010b).
6. Doyle II (2011).
7. Deudney (2007); Doyle II (2010a, 2013, 2015a, 2015d).

Chapter 1

Introduction

This book is concerned with rethinking nuclear ethics as the world's states and peoples continue to suffer under an undeserved and unjust existential burden imposed by a handful of states whose determined reliance on nuclear weapons for their security and status[1] over the past seven decades seems likely to last beyond this century's end. This enduring problem of nuclear existential risk to the world's states and peoples raises questions of the legitimacy of the international nuclear order that privileges the interests of the nuclear-armed few against those of the non-nuclear-armed many. This is to say, the enduring problem of nuclear existential risk is a problem of international nuclear order; and if such an order sustains an undeserved and unjust existential risk, then it seems urgent for nuclear ethics to rethink the question of a just international nuclear order.

Previous attempts to "think" nuclear ethics succeeded in describing and defending an array of competing viewpoints both for and against nuclear defense and deterrence policies, which became one of the pillars of the seven-decade-old international nuclear order.[2] Unfortunately, these previous attempts failed to arrive at a consensus on which moral values or imperatives had lexical priority, that is, could not be rightfully overridden by a more important concern. The failure to arrive at a consensus was tolerable once it became clear that the international nuclear order had become adequately stable, even if not adequately just.[3] However, the second decade of the twenty-first century has witnessed a resurgence of intensified insecurities and instabilities, and, in this context, the lack of consensus in nuclear ethics is no longer tolerable and a renewed effort to rethink nuclear ethics in the context of the international nuclear order must be undertaken.

This chapter is dedicated to setting the stage for this effort at rethinking nuclear ethics for the remaining decades of the twenty-first century. In what

1

follows, the book's general argument will be motivated in more depth. Then, the term "nuclear ethics" will be defined more clearly and distinguished from other modes of normative thought in the study of international security within the discipline of International Relations (IR). Finally, the chapter will preview the book's main lines of argument.

MOTIVATING THE ARGUMENT: QUESTIONS OF SURVIVAL, INTERNATIONAL ORDER, AND INTERNATIONAL JUSTICE

The first two decades of the twenty-first century have been characterized by intensifying national, international, and global insecurities, indicated in part by the fact that nine states continue to steadfastly rely on nuclear weapons for their security and status interests. These states are the United States, Russia, Great Britain, France, China, India, Pakistan, Israel (opaquely), and North Korea. Among these nuclear-weapon states (NWS), the fault lines of distrust and antagonism have deepened most prominently between the United States and Russia, the United States and North Korea, the United States and China, the United States and Iran (which is currently regarded as the most likely nuclear-aspirant state among non-nuclear-weapon states [NNWS]), and India and Pakistan.[4] In its response to these revived antagonisms, the United States has committed $1.5 trillion over the next thirty years to nuclear weapons modernization.[5] Russia and China also have begun nuclear weapons modernization programs, even though their spending commitments cannot match the corresponding U.S. commitment.[6] These increased bilateral tensions have occurred despite long-standing efforts by the international community to develop robust multilateral regimes to prevent new nuclear proliferation, make progress in nuclear disarmament, and foster greater security cooperation among former and present nuclear adversaries. The failure of these security regimes constitutes one of the major failures of the international nuclear order.

In addition to these bilateral tensions, a significant dispute has erupted among several member states of the Nuclear Nonproliferation Treaty (NPT) regime over the nature of the legal and political commitments of the NPT NWS, or N5, to nuclear disarmament.[7] On one side are the N5 and the "umbrella states"—that is, roughly thirty European and East Asian NNWS that shelter underneath the extended nuclear deterrence guarantees of the United States and the nuclear-armed North Atlantic Treaty Organization (NATO). On the other side are the nuclear abolitionist NNWS that believe that the NPT's Article VI makes nuclear disarmament an immediate and ultimate regime priority. This intense dispute indicates one of the major global insecurities of the early twenty-first century—namely, that the very possession of nuclear weapons by the N5 (and other nuclear-armed states) imposes on all other states and peoples a constant and unremitting existential risk of accidental or deliberate nuclear use. Moreover, this N5 and NNWS dispute constitutes a second major failure of the international nuclear order.

The character of this global nuclear existential risk is symbolized by the Doomsday Clock, which the editors of the *Bulletin of the Atomic Scientists* recently set at "two minutes before midnight."[8] In the seventy-five years since the United States used atomic weapons against Hiroshima and Nagasaki to hasten World War II's end, the *Bulletin* has set the Doomsday Clock only three times this close to "midnight": 1953, 2018, and once again in 2019. In 1953, it was set at "two minutes" based on the United States' and former Soviet Union's decisions to test hydrogen weapons, which are much more powerful in destructive capacity than the atomic weapons used against Japan.[9] In 2018 and 2019, it reset the clock to "two minutes" based on the increased nuclear tensions mentioned previously, and especially by U.S. president Donald Trump's initial efforts at nuclear brinksmanship with North Korean leader Kim Jong-un.[10] By contrast, the *Bulletin* moved the clock to twelve minutes before midnight in 1963, after the two superpowers were driven to conclude a series of nuclear arms control agreements in the wake of the Cuban Missile Crisis: for example, the 1963 Limited Test Ban Treaty, the 1967 Outer Space Treaty, and the 1971 Seabed Arms Control Treaty.[11] Almost thirty years later, it moved the clock to seventeen minutes before midnight in 1991, two years after the fall of the Berlin Wall and during the transition of the former Soviet Union to the Russian Federation, which hailed the end of the Cold War. The difference of fifteen minutes of the Clock's settings between 1953 and 1991 symbolized the progress of nuclear arms control and the hope of nuclear disarmament as a peace dividend. These were the hallmarks of the international nuclear order of restraint.[12] Unfortunately, their most recent placement of the Clock's hands back to "two minutes to midnight" signals a radical diminishment of that hope and perhaps the end of progress (at least in the short term) in nuclear arms control and disarmament.[13] For the nuclear abolitionist community, this return to "two minutes to midnight" indicates a global emergency in moral, legal, and political terms, which necessitates immediate action to achieve nuclear disarmament. In terms of the international nuclear order, it indicates a persistent injustice that must be remedied before it is too late.

To undertake the project of rethinking nuclear ethics in the context of "two minutes to midnight," it is important to revisit a core set of questions that have been salient throughout the nuclear age. How should states respond to existential threats, especially if the adversary is nuclear-armed? Which kinds of responses to such threats are morally required? Which are permissible?

Which are impermissible? Moreover, what should states do if the first round of responses to existential threats generates additional existential threats to humanity at large? Additionally, it is important to address questions independent of the occurrence of immediate (nuclear) existential threat. Thus, is it appropriate for the N5 to dominate the politics of nuclear arms control and disarmament agreements? Are nuclear weapons distinct in their moral status—that is, are they uniquely evil weapons? If so, must they be eliminated independent of any other moral, legal, or political consideration? Or do they remain permissible as instruments of deterrence? And if they are permissible as instruments of deterrence, is there something morally significant about the identity of a liberal democratic state (or identity of a liberal democratic alliance or security community) such that nuclear deterrence is impermissible for them?

Taken together, these sets of questions invoke four core values that deserve careful reexamination. The first two values are closely related: *survival* and *security* of the world's states and peoples. These values are distinct insofar as security presupposes survival while survival does not entail security. Despite this distinction, their frequent pairing in the IR literature suggests a widespread scholarly belief that they are practically synonymous. This book treats survival and security as intrinsic moral values: that is, values that cannot be given up and are "ends in themselves." Moreover, these values can be applied across many (and possibly competing) levels of analysis. For questions of nuclear ethics, the three salient levels of analysis include individual persons, states, and humanity at large. As the atomic bombings of Hiroshima and Nagasaki reveal, an extraordinarily large number of individual human lives can be extinguished by a single atomic or nuclear detonation, even if a state that suffers such an attack survives. However, the Doomsday Clock symbolizes the moment in which the security and even survival of states and humanity itself succumbs to accidental or deliberate initiation of nuclear war in the name of national defense and security. Accordingly, one of the most important issues that this book reexamines is the rank ordering of state versus individual or human security in nuclear ethical accounts.

A third and related value that the preceding questions suggest is *international order*. William Walker contends that

> beyond basic survival, the achievement of *order* is—and has to be—the preeminent and perennial concern of states, and especially of the great powers among them, given the existence of this ultimate instrument of destruction and symbol of state power [i.e., nuclear weapons].[14]

For Walker, an enduring condition of state security requires the construction and maintenance of a durable international order. In a nuclear-armed world, this implies a global consensus on a defined set of rules and relationships that enable them to know which nuclear-related policies are necessary, permissible, and impermissible. Walker argues that the world's states have chosen an order of nuclear restraint as the most practical means to ensure world survival since the beginning of the Cold War. And yet, an alternative order of nuclear disarmament promises to secure humanity against the risk of nuclear war by eliminating nuclear weapons themselves. This book reexamines these distinct orders and seeks to determine which is most conducive and morally consistent with the intrinsic values of survival and security in a world characterized by "two minutes to midnight."

A fourth value concept suggested by the preceding questions is *international justice*. Traditionally, justice is defined in moral terms as the condition in which one's reception of benefits and burdens is deserved.[15] Law codifies our moral intuitions about desert within domestic and international political orders, and from this we derive conceptions of civil and criminal justice. The condition of nuclear existential threat qualifies as a grave burden, and thus we might ask if it is undeserved in legal or moral terms. We might also ask if certain responses to nuclear existential threat create additional injustices. Indeed, the question about some liberal democratic states' reliance on nuclear defense and deterrence in the context of liberal commitments to human rights and the rule of international law can be understood as a question of the in/justice of "liberal nuclearism."[16]

In short, the task of rethinking nuclear ethics in the context of "two minutes to midnight" requires a careful reconsideration of the priorities among the values of survival, security, international order, and justice. Accordingly, we might synthesize the various questions related above into one central question: that is, *does the survival and security of states and peoples from nuclear existential threat depend upon the realization of a just international nuclear order?* This reformulated question crystallizes and focuses the book's central concern. It also suggests an analytical structure for examining the whole and parts of the moral problem of nuclear existential threat and the corresponding options for security policy.

For instance, this reformulated question recommends fresh examination of the following issues: in cases of nuclear existential threat or nuclear conflict, who deserves to survive? Is it morally justified to claim that *our* lives—that is, my life and the lives of my fellow compatriots—have greater value than those living in an adversary state? Or, is it rather that our lives have neither greater nor lesser value than their lives? In this regard, how should the current international order be reformed or restructured? If the world's most powerful states (e.g., the N5) determine the international (nuclear) order, and if that order conforms to their national interests, then is it committed to the assumption that some lives have greater value and therefore deserve to survive even at the expense of other (less powerful) lives? If so, does an international nuclear order based on this assumption satisfy the fundamental requirements of international justice? If not, then must we rather assume that all lives have equal value independent of the state in which they reside and that no state or people deserve to lose their security or survival so that another (more power-ful, wealthy, or privileged) state or people might survive and remain secure? In short, must we turn away from exclusive notions of security and toward an inclusive and common security commitment?

In response to the book's reformulated question, the key argument advanced in these pages is that the intrinsic value of the survival and security of the world's states and peoples (and by extension, the survival of humanity itself) from existential harm must not (and conceptually cannot) be over-ridden by competing values. Therefore, an international nuclear order that aims at the common security of all states and peoples must be anchored on fundamental notions of justice. An international nuclear order that privileges the political interests of the most powerful states leads to fundamental and undeserved insecurities for most if not all states and peoples. And such con-ditions cannot but introduce instability into the wider international order and increase the risks of nuclear catastrophe or human extinction. *In short, this book argues that our common survival and security for the indefinite future requires a just international nuclear order.*

Before this argument can be adequately advanced and defended, a few more stage-setting tasks must be completed. In the next section, I begin the process of defining more clearly the role of nuclear ethics in addressing the questions of a just international nuclear order in contrast to other scholarly approaches limited to political, strategic, or legal analysis. Afterward, I offer a more precise definition of nuclear ethics, its epistemological and ontologi-cal assumptions, and the methods to which it is committed.

CONTRASTING NORMATIVE APPROACHES

In the discipline of IR, the dominant modes of policy-relevant normative analysis (as opposed to the narrower social-scientific mission of description and explanation) are often limited to political, strategic, or legal approaches. Political analysis approves or disapproves of nuclear defense or deterrence based on a state's power and status considerations in specific contexts of regional or global threat environments. Such recommendations are anchored solely on the exclusive value of national survival and security.[17] North Korea evidently became convinced of the need to acquire a nuclear deter-rent in order to prevent the United States from engaging in regime change operations.[18] On the other hand, NNWS such as Austria, Brazil, and Ireland forcefully oppose all nuclear defense and deterrence postures based in large part on their assessment of the expected spillover effects of a nuclear conflict between their NWS allies or neighbors.[19] Such NNWS' political preferences thus favor an international nuclear order that is not skewed solely to the interests of the most powerful states.

In comparison, strategic analysis seeks to determine the most effective or efficient policies for implementing political objectives.[20] Thus, if political analysis prescribes nuclear defense and deterrence generally, then strategic analysis determines the quantitative and qualitative elements of a nuclear defense capability sufficient for that need. Accordingly, U.S. presidential administrations in the post–Cold War period have routinely published a "nuclear posture review" or "national security strategy" to explain the overall role of nuclear weapons in its deterrence and defense policy.[21] At the level of military high command, a nuclear war plan details the strategically appropriate uses of nuclear weapons across a wide array of anticipated contingencies. It simultaneously expresses the two elements required for credible deterrence postures: capabilities and intention. This is to say, the existence of a nuclear war plan signals the possession of nuclear weapons and the leadership's intention to engage in nuclear warfare under certain conditions. Of course, states whose political interests compel nuclear avoidance will not advance a strategic plan that incorporates nuclear weapons. However, some NNWS have chosen to shelter under the U.S. extended nuclear deterrence guarantee, that is, the "nuclear umbrella."[22] Their choice compels policy decisions on questions of nuclear sharing arrangements or the permissibility of allowing U.S. nuclear submarines to dock in port. Obviously, the abolitionist NNWS' political commitments rule out any strategic plan incorporating nuclear defense or deterrence. In the end, it seems clear that strategic analysis has little to say about the necessity of just international orders for the realization of humanity's survival, unless it includes a plan for how to reduce and eventually eliminate nuclear weapons in a safe and verifiable manner.

The third kind of normative IR approach is international law. We might expect that a state's legal stance on nuclear defense and deterrence maps onto its political stance, which would lend credence to the claim that international law can often function as a mode of power politics.[23] For instance, the N5 have benefitted from the 1996 International Court of Justice's (ICJ's) Advisory Opinion on the Legality of Nuclear Weapons, which stated that nuclear defense against (nuclear) existential aggression was not clearly prohibited by international law.[24] In contrast to the ICJ's opinion, the abolitionist NNWS have emphasized the N5's binding legal nuclear disarmament obligations as expressed by the NPT's Article VI and as clarified by the 2000 and 2010 NPT Review Conference Final Reports.[25] Additionally, the abolitionist NNWS hold that the 2017 Treaty on the Prohibition of Nuclear Weapons (TPNW) removes the justificatory cover that the ICJ's 1996 advisory opinion provided for the N5's reliance on nuclear defense and deterrence.[26] Even so, the N5's refusal to recognize the TPNW (see chapters 3 and 5) means that the international legal status of nuclear weapons remains indeterminate.

The upshot of this discussion of normative IR approaches is that an ethical approach to the questions of the necessity, permissibility, or impermissibility of nuclear defense and deterrence (as well as other nuclear issues) must focus on the priority of values independent of (or not reducible to) strictly political, strategic, or legal values: which is to say that these questions must be approached by way of established ethical or moral theories. The tasks of the next section are to situate nuclear ethics as an academic subfield, describe its epistemic and ontological commitments, and then examine more thoroughly how nuclear ethics applies ethical theory—that is, the methodology of nuclear ethics—to render judgments on the necessity or im/permissibility of nuclear policy.

NUCLEAR ETHICS: EPISTEMIC COMMITMENTS, ONTOLOGICAL COMMITMENTS, AND METHODOLOGY

Nuclear ethics is an interdisciplinary inquiry that applies moral or ethical theory to the nuclear-related questions of international security and world order.[27] In the discipline of moral philosophy, theories of value describe, explain, and justify the conditions of moral requirement, permissibility, and impermissibility. As such, these theories are anchored to a diverse set of metaethical commitments.[28] Metaethics is concerned with establishing if ethics or morality is a system of *true* statements concerning intrinsic and instrumental values. If ethical statements are true, then the character of moral reality is accurately revealed on questions such as the universality of value, moral rights and duties, or the moral *summum bonum* or *summum malum*. As chapters 2 and 3 suggest, contributors to the nuclear ethics debate intend to do more than relate plausible normative inferences on nuclear defense and deterrence; indeed, they intend to relate true normative statements or moral facts on states' nuclear commitments.

Accordingly, it is important to note the controversy among philosophers concerning the possibility of moral statements possessing truth-value—that is, expressing moral facts—as opposed to expressing solely an actor's desire, preference, or interest. The former stance is called moral realism and the latter stance is called emotivism or noncognitivism.[29] It is not necessary to trace here all or most of the philosophical debate between these two stances. However, it is interesting that IR Realism, inasmuch as it contributes to normative IR theory, is unequivocally committed to emotivism insofar as it reduces questions of ethical value to politics and thus to questions of interest and preference.[30] Thus, mainstream IR approaches not only describe or explain state behavior solely in terms of interest, but they prescribe courses of action consistent with interest while maintaining a public face of moral skepticism.[31] Emotivism and moral skepticism are contradictory stances, and given the intrinsic character of the survival and security values invoked in questions of nuclear defense and deterrence, it is arguably important for nuclear ethics to adopt a metaethical stance that does not suffer from contradictions of this kind.

For these reasons, this book assumes the stance of *moral realism*.[32] It acknowledges that there is often passionate disagreement about moral claims; nevertheless, it insists that disagreement is not sufficient reason to reject the possibility of moral facts. Rather, it insists that disagreement presupposes their possibility. Moral realism also acknowledges that claims of moral truth are linked in important ways to actor's motivations, desires, preferences, and interests; even so, actors can misunderstand their interests and, hence, draw mistaken normative inferences. Thus, for the moral realist, it must be possible to determine the truth-value of statements such as "nuclear hostage holding is wrong" or "nuclear deterrence is a bloodless strategy."[33] And just as the truth-value of an empirical claim is based on its correspondence with the real world, the truth-value of a moral statement is based on its correspondence with value-laden social and institutional realities that have been constructed and modified by collective human agency.[34] In the end, if moral realism is correct, it follows that we can know (at least in principle) what the *summum bonum* or *summum malum* is with respect to questions of nuclear defense, deterrence, or disarmament.

Having now identified the metaethical commitments in moral realism, it is now important to distinguish between three kinds of ethical or moral theories from which this book's applications will come. One ethical theory is consequentialism. Its main imperative prescribes the achievement of good outcomes and the avoidance of evil outcomes. For consequentialists, action in itself is neither intrinsically good nor evil, since the moral quality of action is not taken to be independent of the outcomes it produces. If all social or political action is a means to an end, then the only measure of its rightness is the achievement of good outcomes. To the question "good for whom," consequentialism distinguishes between narrower nationalist/statist and cosmopolitan approaches. The former counts only fellow compatriots as morally relevant subjects while the latter includes all affected persons of action independent of national identity. Accordingly, the former approach understands "greater good" only in terms of nationality and is willing to privilege the national good over the global good, while the cosmopolitan privileges the global good over any narrower national good if the two cannot be harmonized. Based on this contrast, only cosmopolitan consequentialism truly counts as "utilitarian," even if nationalists mistakenly describe their reasoning with that term.[35] One implication of this contrast among consequentialist theories is that moral dilemmas arise if there is an absence of a metaethical principle to resolve conflicts between competing nationalist or cosmopolitan imperatives. As the subsequent chapters discuss, this is often evident in the nationalist-utilitarian debate over nuclear defense and deterrence in which the former defends such policies against the latter's rejection of them.

Deontological ethics is a contrasting moral theoretic approach from which nuclear ethics might find useful applications. The Greek term "deon" or "deont" refers to the concept of duty[36] that functions to define the rightness or wrongness of an action within determinate social contexts. On this view, the general duties that each agent bears are rooted in their humanity while specific duties are rooted in the acquired or assigned social role(s) that they possess.[37] Accordingly, each human being bears an intrinsic right to life and security. Each human being also bears a corresponding general duty to preserve every other human being's right to life and security within their sphere of activity to the best of their ability. Thus, acting in defense of one's own or another's life is morally right even if one's defense fails. Moreover, acting in self-defense or the defense of another is morally right even if it means wounding or, in some cases, killing those who have mounted an attack. Persons who hold social roles such as police officer or soldier bear specific duties in the protection of life, which they must execute even if they fail or lose their own lives in the process.

As with consequentialism, deontology can admit of diverse applications that might impose moral dilemmas on political actors. Thus, we might expect that a president or prime minister's official duties to implement nuclear defense or deterrence policy consistent with the right of national self-defense in international law could conflict with their other official and personal moral duties to humanity in contexts where an imminent threat appears on their horizon. The task of deontological analysis is to determine if a leader's conflict of moral duties can be reconciled and, if so, in whose favor. In just war theory, the national self-defense requirement is a statist application of deontological ethics, and it directs presidents or prime ministers to prioritize state security above all else.[38] Alternately, for Kantian theory, the imperatives of justice and cosmopolitan peace combine to direct such leaders to rank human security as equally important with national security in all except the most extreme cases.[39]

A third ethical or moral theory is virtue-ethics. It emphasizes the moral imperative to express or exhibit in one's action the elevated virtues of character, such as courage, humility, temperance, or prudence.[40] Virtue-ethics does not associate morally praiseworthy action with obedience to moral duty nor with the realization of good outcomes for the few or the many. Rather, on this view, morally praiseworthy action evokes traits associated with divine nobility or human excellence. Thus, we honor the soldier who fought bravely to save the lives of her comrades-in-arms, even if she ultimately loses her life. Similarly, we honor the perseverance of the head of state or diplomat who tirelessly pursues a durable and just peace among enemies even if their pursuit fails. By contrast, we find an official's betrayal of comrades for naked self-interest ignoble and shameful. This trio of examples in the context of the distinctions made among virtue-ethics, consequentialism, and deontology suggest that a virtue-ethical analysis of nuclear issues is less concerned about the consequentialist balance between nuclear risks and rewards[41] or if citizens of nuclear-armed states have given consent to become nuclear hostages.[42] Instead, it is more concerned about the lack of moral courage to confront conventional nuclear wisdoms that suffer from significant political, strategic, or legal blind spots and that has led all relevant actors to lose sight of their moral duties and then fail to achieve the greater moral good.[43] If such a virtue-ethical analysis is compelling, then there is even stronger reason to undertake a rethinking of nuclear ethics in the context of "two minutes to midnight."

Enough has been said above concerning the range of relevant ethical or moral theories to permit us to turn to the "applied" element of nuclear ethics. As chapters 2–5 demonstrate, the nuclear ethics literature is comprised of single-theory and hybrid theoretic applications. Typically, IR realist or rationalist analyses are limited to consequentialist applications and Kantian analyses are limited to deontological applications. By contrast, the just war theoretic tradition is hybrid insofar as it includes deontological and consequentialist imperatives in the determination of "just cause" or *jus ad bellum* and "just means" or *jus in bello*.[44] In addition, one of the most prominent hybrid applications from the later Cold War era is Joseph Nye's "three-dimensional" nuclear ethics.[45] Nye's account begins by identifying the logical possibilities for moral assessment if the categories of motive (virtue), means (duty), and consequences are limited to "good" and "bad" value assignments. Table 1.1 relates these logical possibilities below.

Nye's model effectively captures the breadth of our moral concerns, and in doing so, it urges against the myopia of single-theory approaches. Of course, two possible virtues of a single-theory approach is its simplicity and clarity in the application of moral decision and assessment. However, Nye correctly argues that single-theory approaches prevent due consideration of excluded moral concerns or viewpoints that must be included for moral choice and assessment to count as responsible.[46] For instance, moral action undertaken solely on the basis of virtuous motives ignores the negative moral implications of actions that violate important rights or duties or that produce harmful consequences. Similar sorts of unacceptable trade-offs follow for myopic concerns to act dutifully or produce good

Table 1.1 Nye's Three-Dimensional Model of Ethical Possibilities[i]

Case	Motive (Virtue)	Means (Duty)	Consequences (Outcome)
1	Good	Good	Good
2	Good	Bad	Good
3	Good	Good	Bad
4	Good	Bad	Bad
5	Bad	Good	Good
6	Bad	Bad	Good
7	Bad	Good	Bad
8	Bad	Bad	Bad

[i] (Nye 1986, 22).

outcomes. By adopting a hybrid approach as related by Table 1.1, moral vision expands and becomes appropriately sensitive to the breadth of moral concern. Such an approach rightly trades off the false hope of moral simplicity for the reality of moral complexity in the kinds of cases that nuclear ethics is most likely to engage.

To this end, careful consideration of Table 1.1 indicates that cases 1 and 8, respectively, are unambiguous instances of moral and immoral policies. Case 1 refers to instances where morally upright motives and actions are expressed and where morally good consequences ensue. This is the ideal case that sets the standard for ethically informed policymaking. Case 8 is the categorical opposite of case 1, and therefore, it represents options or scenarios that must be absolutely avoided or condemned. By contrast, cases 2–7 represent the broad range of nonideal instances of moral choice and assessment. Each case has at least one negative assignment combined with at least one positive assignment across the three value categories. And since Nye rightly believes ethical myopia is irresponsible, the crucial methodological question for a three-dimensional nuclear ethics is the proper weighting of each value category in relation to the other two.

One possible but odd approach to the proper weighting of moral values might seek the simplicity of numerical advantage: two good categories outweigh one bad category, and two bad categories outweigh one good category, irrespective of the values at issue. On this approach, case 2 would suggest that nuclear deterrence is morally justifiable because it is well intentioned and it has led to good outcomes (war prevention) despite the deontological moral fact that threatening innocent persons in enemy states with nuclear death is wrong (i.e., murderous). While some commentators covered in chapters 2–5 seem to accept this kind of numerical weighting, a serious problem emerges if it is applied to other cases. For instance, in case 3 we find nuclear deterrence is well intentioned and counts as a good act because it is aimed at producing the good outcome of war prevention. However, deterrence failure leading to nuclear catastrophe must count as a "bad" outcome. Are we truly willing to say that nuclear deterrence is morally preferable to conventional deterrence or conflict resolution in such a case? And yet, if two "goods" override one "bad" on a simplistic numerical analysis, then we would be committed absurdly to the moral justification of nuclear deterrence even if it led to nuclear apocalypse.

An alternative approach to value weighting is to privilege only one of the categories as having greater *prima facie* moral authority. The virtue of this method is that it avoids the absurdities of the aforementioned approach. Unfortunately, it might also reproduce the vices of the single-theory method if undue respect is paid to one value category at the expense of the others. For instance, chapters 2 and 3 identify some just war theorists as acknowledging the salience of motive, means, and consequences with respect to nuclear defense and deterrence; even so, by privileging the *jus ad bellum* right of national self-defense of Western states in the name of preserving freedom of religion, it seems that they are always willing to accept an undue risk to the survival of humanity.[47]

My interpretation of Nye's nuclear ethics is that it expresses this rule-consequentialist sentiment. Rule consequentialism "selects rules solely in terms of the goodness of their consequences and then claims that these rules determine which kinds of acts are morally wrong."[48] This is to say, the guiding imperative of rule consequentialism is to give preference to rules and duties that have the greatest likelihood of avoiding evil outcomes and, hopefully, producing good outcomes. Accordingly, when discussing his approach to moral assessment that expresses integrity and responsibility, Nye states, "Moral integrity would not mean locating a point of equal distance or equilibrium between means and consequences, but a disposition towards certain qualities in the moral reasoning used in making trade-offs."[49] He describes this disposition as, first and foremost, a matter of maintaining a "strong presumption in favor of rules and rights."[50] This remark might suggest that Nye intends to weigh deontological values somewhat higher than consequentialist values. However, he later discusses the importance of treating rules as "prima facie" duties that might require suspension if the need for the greater good must be satisfied. Indeed, Nye supports this orientation by stating that a "sophisticated consequentialist analysis" concentrates on the question: "if I override normal moral rules because it will lead to better consequences in this case, will I be damaging the institution by eroding moral rules in a manner which will lead to worse consequences in future cases?"[51] This question suggests that the "strong presumption in favor of rules and rights" is a means of securing "better" rather than "worse" consequences, which is what we expect from rule consequentialist approaches.

If the preceding discussion is correct, then it explains Nye's defense of nuclear deterrence as an instance of case 2 in Table 1.1. This is to say, Nye's overall judgment is that nuclear deterrence's positive record at preventing superpower nuclear war (good consequence) along with the ongoing desire to prevent superpower nuclear war (good motive) constitute the greater moral value even though advancing nuclear threats increase the risks of human extinction (bad means).[52] Alternately, it also explains his concern against counter-city nuclear deterrence on the basis of the duties that we owe to innocent people, which seems to me to count as an instance of case 4 or, given the tradition of nuclear nonuse, case 6.[53]

A third possible methodological approach to value weighting is to determine the weights and assessments of competing values purely on a case-by-case basis. To the best of my knowledge, commentators to the nuclear ethics literature do not favor this approach. However, as a methodological possibility, its virtues might include an increased sensitivity to the nuances of each case as well as a lack of dogmatic commitments to any specific ethical theory. On the other hand, this approach might also be prone to gross moral inconsistency, partiality, or contradiction across a wide range of moral choices and assessments. If so, then its main vice would be a wooliness or similarity to pseudo-scientific cousins such as astrology. Perhaps the closest that the literature on nuclear defense and deterrence gets to this third methodological approach is the analysis of cases in which it is claimed that U.S. nuclear deterrence is morally justified but that the nuclear deterrence policies of U.S. adversaries are not (e.g., North Korea) even though each enjoy the presumptive motive of *jus ad bellum* right of national self-defense and even though their combined deterrence postures have resulted (so far) in war prevention (i.e., case 2). I say "perhaps" because this is often portrayed as "objective" foreign policy analysis free of moral presumption even though it "smuggles in" the moral values consistent with a narrow nationalism.[54] Only occasionally are the moral assumptions and arguments of such views made explicitly.[55]

Having now described nuclear ethics as a species of applied ethics with moral realist commitments, it is time now to relate this book's methodological commitments. First, this book adopts a hybrid approach and it regards single-theory approaches as myopic and overly simplistic. Accordingly, it accepts that questions of nuclear defense, deterrence, and disarmament are often subject to difficult if not intractable moral dilemmas, although it argues that their intractability is primarily a function of the security conception on which state adversaries rely.[56] This element of the book's argument will be fleshed out directly in chapters 3 and 6.

Second, in contrast to Nye's rule consequentialist approach, this book regards the question of human survival and the security of states and peoples as matters of intrinsic moral value and therefore necessarily as matters of cosmopolitan moral duty and right. More specifically, it takes individual and collective survival and security as having equal moral value, insofar as individual persons are social creatures whose survival and security are inextricably intertwined with their communities of reference, or what Michael Walzer and John Rawls refer to as "peoples."[57] By taking the survival and security of individuals, peoples, states, and humanity as equally weighted deontological values, this book's core premise is that, in the context of "two minutes to midnight," it is a cardinal error of moral reasoning to weigh the intrinsic survival right of any one state or people over another and of any one state/people or collection of states/peoples over humanity generally. These are cardinal errors precisely because each individual and their communities of reference possess the status of moral ends in equal measure.[58] Thus, one responsibility of a nuclear ethics for the twenty-first century stuck at "two-minutes to midnight" is to rethink the assumptions of earlier nuclear ethical analyses that, on the one hand, awarded greater moral value to one's own national survival or security or that, on the other hand, awarded greater moral value to humanity at the expense of the survival/security rights of states under certain conditions. The product of such rethinking is a moral reconceptualization of the security imperative, which this book refers to as *common security*. This effort at reconceptualization is assigned most directly to chapters 3 and 6.

Third, as suggested immediately above, rethinking nuclear ethics in the context of "two minutes to midnight" involves assessing the need for the reconceptualization of key ethical concepts in light of the world's collective experience with nuclear security dilemmas. Conceptual analysis as a methodology begins by problematizing the uses of one or more key terms as they arise in discourse or argument. Thereafter, these terms are subjected to a comparative case study analysis in which its varied uses are traced through other relevant or salient discourses or arguments. The purpose of this comparative case study is to determine patterns or degrees of acceptable usage and any corresponding conceptual entailments.[59] For instance, the previous paragraph suggested that the concept of security is a deontological value possessed in common by all individuals, peoples, and states. However, it has been acknowledged that many, if not most, commentators use "security" in an exclusive nationalistic sense. If this book's conceptual analysis of "security" is correct, then any exclusive conception of national security is morally problematic if it is taken to override common security. And in regard to this book's central question, the broader mission of nuclear ethics in the context of "two minutes to midnight" is to rethink the concepts of international order and international justice consistent with the common security conception.

Finally, the task of rethinking nuclear ethics must continue to guard against partiality. Nye defines impartiality as having a respect for the interests of others.[60] Unfortunately, this definition does not capture the degree to which "having a respect for" and "being partial to" can coexist if one's ethical analysis is committed to an exclusive national security conception. For instance, it seems common for nationalist advocates of nuclear defense and deterrence to "respect" the survival and security interests of rivals while regarding their own survival/security rights as overriding. In this limited sense, "respect" is nothing more than acknowledgment or perhaps appreciation of an adversary's predicament as opposed to an acceptance of an other-regarding duty in the context of conflicting interests. The upshot is that ethical argument regarding nuclear defense, deterrence, and disarmament must, if it is impartial in the strictest sense of the term, give equal moral regard to self, allies, and adversaries alike. It then follows that impartiality is necessary for practices of ethical justification insofar as it must avoid the imposition of undue double standards. For instance, an impartial justification of nuclear deterrence as instrumental for the good of human survival would have to admit, *ceteris paribus*, that each state that seeks nuclear deterrence is justified in doing so, even if one or more of them are U.S. adversaries. And if it is false or unjustified that nuclear deterrence serves the good of human survival, then we might be compelled to accept that nuclear deterrence is immoral universally.

Having motivated the book's inquiry, related its assumptions and commitments, and identified the key elements of its methodology, the remainder of this chapter previews the book's arguments. Attention will be given to the whole of the argument as well as how its parts are distributed across the five chapters that follow.

A PREVIEW OF THE ARGUMENTS

It is useful to begin this preview by recalling the book's reformulated question: does the survival and security of states and peoples from nuclear existential threat depend upon the realization of a just international order? This question suggests at the very minimum two possible responses. One is that the survival and security of states and peoples requires a stable international nuclear order even if it does not satisfy the basic requirements of international justice, and the other is the requirement of a minimally just international nuclear order. The latter response requires an account of minimal international justice in ways relevant to the existing international nuclear order even if it cannot or ought not to be applied to other facets of the world order, such as the international economic order. Accordingly, the chapters sometimes refer to the former response by the shorthand expression "Survival > Order" and to the latter by "Survival > Order > Justice" or "Survival > Just Order."

The book's fundamental contention is that the survival and security of states and peoples requires a minimally or fundamentally just international nuclear order. To support this contention, the chapters advance several propositions. The *first* is that, during the Cold War (chapter 2) and post–Cold War eras (chapters 3 and 5), the debate on the moral justifiability of nuclear defense and deterrence failed to reach consensus on the normative priority of national survival and security in contrast to the priority of human survival and security. On one side, it was argued that national survival and security were just causes that could not be overridden by any another competing normative requirement. Interestingly, some liberal commentators agreed with this overtly nationalist and "realist" claim as it coincided with the value of the survival of liberal democratic orders against hostile and nuclear-armed illiberal and totalitarian states (chapter 4). Accordingly, this ultimate national (and liberal) survival right is the ground for the moral justification of nuclear deterrence to prevent nuclear aggression or any other kind of imminent existential menace. On this view, the need to prevent nuclear aggression, or a conventional conflict that might lead to nuclear use, led to an international nuclear order in which deterrence and (later) strategic stability took precedence over any (perceived) violations of arms control and disarmament agreements or related promises that could fuel claims of injustice by some states-parties. Moreover, each NWS in such an international nuclear order threatened to resort to nuclear defensive action (i.e., nuclear reprisal strikes) if it could not prevent nuclear or catastrophic conventional aggression against its territory or interests. In such tragic cases, nuclear defense was taken as morally necessary to punish the aggressors, secure the surviving elements of the victimized state, and then restore nuclear deterrence and stability in the international nuclear order.

On the other side, it was argued that human survival and security are ultimate values that could not be overridden by competing nationalist imperatives of state survival and security. Two kinds of supporting propositions were offered in this regard. One focused on the priority of fundamental human rights of survival and security over the derivative rights of states (i.e., governments or national regimes) to their continued existence. A second kind focused on the overriding imperative of the greater global good against competing nationalist claims. Both kinds of supporting propositions express a moral cosmopolitanism that encompasses deontological and utilitarian imperatives, respectively. Taken together, they rule out the permissibility of any offensive and defensive uses of nuclear weapons that would kill thousands if not millions of innocent human beings and where countless others would suffer grave injuries merely to ensure the survival of one or a few governments. Additionally, on this view, even if nuclear weapons were never used in warfare, it would still be morally unconscionable for any NWS to threaten their use against an adversary state and put its population at risk of nuclear death. Such existential risks are fundamentally undeserved and unjust. It follows on this view that the only morally responsible course of action for the world's NWS and their allies is to pursue and realize without delay a complete, irreversible, and verifiable elimination of nuclear weapons. Accordingly, in contrast to the former position, the international nuclear order must be constructed around the imperative of nuclear abolition and not merely war prevention. An international order so constituted would place the human survival as the ultimate just cause, which leads to the conclusion that an international nuclear order of disarmament is a just international order.

Chapter 2 traces these competing lines of nuclear ethical debate among prominent Cold War–era commentators and explains their lack of consensus (or dissensus) as a function of a specific metaethical problem: that is, the absence of a clearly articulated ethical principle that harmonizes the conflicting imperatives of state and human survival/security *or* demonstrates definitively the priority of one over the other in a bipolar world characterized largely by mutually assured destruction. The first part of chapter 3 traces the continuation of this deadlocked debate into the post–Cold War era even as the basic structure of international politics changed from U.S./Soviet bipolarity to U.S. unipolarity. Additionally, chapter 4 explores one facet of this debate by proponents of liberal democracy that paradoxically found common cause with the advocates of national security and nuclear defense and deterrence. It critically examines the position taken by John Rawls, who was perhaps the most important twentieth-century English-speaking political philosopher, who argued in favor of nuclear deterrence as a justifiable means of preserving democratic ways of governance and social cooperation from nuclear aggression or blackmail arising from irreconcilably hostile illiberal or "outlaw" states. Chapter 4 calls this Rawlsian position "liberal nuclearism." This position strongly suggests that the survival and security of liberal democratic states is more important than that of illiberal states (and their unfortunate citizenries) if the latter seems intent on aggression that might lead to or involve nuclear use. Similarly, it holds that an international nuclear order constructed around war prevention is necessary for liberal survival, even if a just nuclear order is an admirable ideal toward which efforts might be made. Chapter 5 completes the examination of the deadlock in the post–Cold War nuclear ethical debate by critically examining the intense disputes over the N5's disarmament commitments as expressed in NPT Article VI and further defined by the 2000 and 2010 NPT Review Conference Final Reports. It takes note of the increased polarization among N5 and NNWS on nuclear abolition, and it assesses the rise of the Humanitarian Imperative to Abolish Nuclear Weapons as an expression of the human security approach to the international nuclear order.

A *second* proposition in support of the book's main contention is that the deadlock in the nuclear ethical debate can be transcended or resolved if a common security imperative becomes the core anchoring moral principle on which ethical and policy analysis can proceed. Chapter 3 defines common security as a conception requiring the survival and security of all states and peoples in equal measure. Common security thus redefines the core relational dynamic of security by the preposition "with" as opposed to "against." This key distinction between "security with" and "security against" one's adversaries has significant implications for the moral arguments anchored on national and human security conceptions, both of which are classified in chapters 3 and 6 as "security against" conceptions. This latter point deserves some elaboration.

It has been claimed that moral justifications of nuclear defense and deterrence were more often rooted in the normative priority of national (or alliance or collective) security while moral condemnations were more often anchored on the normative priority of human security. Of course, alliance or collective security conceptions are broader in scope than an exclusive national security conception, insofar as each alliance or regime member belongs to a larger unit of the "we" which bears a fundamental survival/security right. As such, alliance and collective security count as cooperative security conceptions. Even so, they are aligned in framing the security relation as "against" external or even internal foes. They assume fatalistically that indefinitely long security cooperation, even among allies, is nearly impossible and that the only constant in world politics is security competition. Thus, the United States (national security) along with Japan and South Korea (alliance security) and the NATO (collective security) are required in moral, legal, and political terms to "secure against" nuclear threats from one or more state adversaries (e.g., Russia, China, North Korea). On the other hand, the state signatories to the TPNW envision the world's peoples and humanity generally as requiring "security against" the morally blind or callous governments of the NWS, each of whom are willing to risk human survival/security for their own preservation. In their view, intractable NWS' security dilemmas and vulnerabilities, important though they may be, will never be eliminated if their exclusive security imperatives are not given up. Consequently, these narrower security conceptions must not continue to hold the world's people hostage to the increasing prospects of nuclear catastrophe. In the end, neither side of the nuclear ethical debate so far has achieved a clear "victory" despite the changes in the structure and power relations of world politics between the Cold War and post–Cold War eras.

This background of scholarly and policy-expert dissensus on nuclear ethics motivates chapter 3's contention that the introduction of common security as an anchoring moral principle provides a promising alternative ethical posture. One feature of this alternative is that "security with" reframes adversarial NWS relationships into cooperative relationships in which the survival and security of each NWS (or nuclear-aspirant state) is mutually implicated. This means that, from the moral viewpoint, the just cause of state survival/security and the just cause of human survival/security are not subject to a zero-sum dynamic entailed by of the competing "security against" conceptions and practices. Since individuals, peoples, and states as a whole deserve survival and security in equal measure, the justice due to each of these kinds of actors can be satisfied in common and without diminishment to any of them. Put slightly differently: if national security is a public good among distinct citizenries, and if alliance or collective security is a public good for all relevant states-parties, then common security is *as a matter of moral fact* a public good for all states and peoples that inhabit the world. Of course, the introduction of "security with" conceptions and practices must necessarily occur while the competing NWS address their respective security dilemmas. However, neither the anarchy condition nor the concrete features of their respective security dilemmas constitute an insuperable barrier to sustained and effective efforts to redefine the relevant relationships away from "enemy" to "rival" or even "friend." "Security with" conceptions and practices can ultimately make irrelevant the resort to nuclear defense and deterrence in ways that comport with rational political analysis.

The *third* proposition in support of the book's main contention is that the common security imperative must generate a set of new nuclear ethical maxims. This proposition is developed in chapter 6. The general thrust of a new set of nuclear ethical maxims is to privilege fundamental elements of international justice defined by the moral duty to seek and realize the mutual security of individuals, peoples, and states. It is expected (although it is not necessary) that the pursuit of common security would begin with NWS adversaries. Afterward, it must include actions to realize common security relationships among (former) NWS and NNWS adversaries and their corresponding peoples. Such transformed relationships would become indicators of a just international nuclear order that, in relationship to the survival and security of humanity, is a necessary condition. It is in this sense that a new set of nuclear ethical maxims would be consistent with the expressions Survival > International Justice > International Order or Survival > Just Orders.

One final remark is important before chapter 2's critical review of the Cold War–era nuclear ethical debate commences: the aim of this book is to advance an argument about *why* a twenty-first-century nuclear ethics should be anchored on a common security imperative and not *how* the maxims related in chapter 6 ought to be translated into policy and action. The latter endeavor is a crucial security studies or public policy undertaking that would likely double the length of this book, and this discussion on how to foster and sustain common security dynamics deserves its own book-length treatment. Additionally, it seems to me that many recent contributions to the empathy, trust, and diplomatic literatures in IR offer promising policy-related inquiries that are consistent with the applied ethical account defended in this book.[61] My hope is that this book's contribution will advance the nuclear ethical debate significantly and, in relation to these other IR literatures, make even more explicit the normative motivations upon which the processes of empathy and trust among states might find their justifications.

NOTES

1. Arms Control Association (2018).
2. Walker (2012, chapter 1).
3. Walker (2012, chapters 3–6).
4. For the conflicts involving the United States, for example, see Roberts (2016). For the India–Pakistan conflict, see Ganguly and Kapur (2010); Sagan and Waltz (2003).
5. Sanger and Broad (2017).
6. Gady (2018); Kulacki (2018).
7. The N5 are comprised of the United States, Russia, Great Britain, France, and China.
8. BBC News (2019); Bromwich (2017).
9. BBC News (2019).
10. Baker and Choe (2017).
11. Arms Control Association (n.d.).
12. Walker (2012).
13. Tannenwald (2018).
14. Walker (2012, 2). Emphasis in the original.
15. Kraut (1992, 319–27). For a comparative discussion of the concept of justice from ancient to modern times, see, for example, Santas (2001). For a discussion on desert, see Feldman and Skow (2015).
16. Doyle II (2015a).
17. Waltz (2003).
18. Baker and Choe (2017); Lankov (2016); Ward (2018).
19. See, for example, Austria (2018); Brazil (2018); Ireland (2018); The New Agenda Coalition (2017).
20. For a comprehensive discussion of the relationship between political and strategic thought, see Gray (1999).
21. See, for example, The White House (2002, 2015); U.S. Department of Defense (2010).
22. See, for example, Morgan (2003, 15–19).
23. Hoffmann (2017).
24. International Court of Justice (1999).
25. 2000 Review Conference of the Parties to the Treaty on the Non-Proliferation of Nuclear Weapons, (2000) (2010 Review Conference of the Parties to the Treaty on the Non-Proliferation of Nuclear Weapons, "Final Document, Voume 1, Part 1" 2010). For a representative statement on Article VI obligations by abolitionist

NNWS, see New Zealand on behalf of the New Agenda Coalition (Brazil, Egypt, Ireland, Mexico, New Zealand and South Africa) (2018).

26. United Nations General Assembly (2017).

27. Doyle II (2010a).

28. Dittmer (n.d.).

29. Dittmer (n.d.).

30. See, for example, Donnelly (1992); Martin (2004).

31. Nye (1986, 7).

32. This discussion of moral realism relies on Sayre-McCord (2017).

33. These "quotes" are paraphrases of, respectively, Lee (1985) and Walzer (2015 (1977), 270–71).

34. For the link to socially constructed reality and normative theory, see, for example, Searle (2010); Wendt (1999).

35. See, for example, Ellis (1992); Martin (2004).

36. Doyle II (2017b).

37. See, for example, Korsgaard (1996, 106–32); Searle (2010, 145–73).

38. Orend (2013, 33–70).

39. See, for example, Hoffe (2006); Ion (2012).

40. See, for example, Coll (1999).

41. Nye (1986, 62–72).

42. Lee (1985).

43. Craig and Ruzicka (2013); Pelopidas (2015).

44. Orend (2013, 33–184).

45. Nye (1986, 22).

46. Nye (1986, 24–26).

47. See, for example, Payne and Payne (1987); Quinlan (2009).

48. Hooker (2016).

49. Nye (1986, 23).

50. Nye (1986, 22).

51. Nye (1986, 25).

52. Nye (1986, 59–80).

53. Nye (1986, 27–41, 108–15).

54. Nye (1986, 7). For one recent instance of such analysis, see John Bolton's critique of North Korean deterrence postures in Gehrke (2019).

55. See, for example, Payne and Payne (1987). This work will receive more detailed coverage in chapters 2 and 3.

56. For a book-length treatment on the moral dilemmas of nuclear proliferation and arms control, see Doyle II (2015b).

57. Rawls (1999); Walzer (2015 (1977), 53–57).

58. Kant (1996, 80–81); Korsgaard (1996, 106–32).

59. Brandom (2000, 45–96); Wilson (1963, 1–48).

60. Nye (1986, 22).

61. See, for example, Holmes (2018); Wheeler (2018).

Chapter 2

Nuclear Ethics during the Cold War

Competing Imperatives and Unresolved Debates

Both the just-war teaching and non-violence are confronted with a unique challenge by nuclear warfare. This must be the starting point of any further moral reflection: nuclear weapons particularly and nuclear warfare as it is planned today, raise new moral questions.

—National Conference of U.S. Catholic Bishops, 1983[1]

Nuclear weapons explode the theory of just war.

—Michael Walzer, 1977[2]

The advent of atomic weapons between 1939 and mid-1945, their use by the United States against the Japanese cities of Hiroshima and Nagasaki in August 1945, and the subsequent development of thermonuclear weapons in the 1950s ushered in a new technological capability of mass destruction that previously had been only remotely conceivable. In 1795, Immanuel Kant had urged states to avoid wars of extermination that "would let perpetual peace come about only in the vast graveyard of the human race."[3] The remote possibility of eighteenth-century wars of extermination was raised by the prospect of enduring ideological warfare motivated by rising popular and democratic sentiments against monarchism. However, the advent of the nuclear age radically transformed the scope of military technology to obliterate states and peoples and, potentially, all of humanity.[4] Thus, the U.S. Catholic Bishops, committed to the Scriptural commandment to care for humanity and its earthly home, expressed shock at the unprecedented moral challenge forced upon humanity by the nuclear age.[5] The secular theorist, Michael Walzer, went so far as to suggest that the morality of warfare itself was obliterated by the nuclear age.[6] To obliterate the morality of warfare is to erase altogether the conception of justice in warfare. Such obliteration also incapacitates any domestic or international order that might organize states and their conflicts toward principles of justice or, at a minimum, a tolerable but unjust stability. In short, the nuclear age made and continues to make practically possible a moment that human society and human life itself faces a day of reckoning from which it might never return. And, as Sisela Bok noted in her reflections on Kant's *Perpetual Peace*, the key challenge in the nuclear age was to organize state behavior during periods of war and cease-fire such that "a day of reckoning never comes."[7]

Unfortunately, since August 1945, this day of reckoning has drawn close more than once. Even before the first uses of atomic weapons, J. Robert Oppenheimer[8] believed that the British practice of conventional oblitera- tion bombing during World War II induced a moral numbness and indiffer- ence within the West over the loss of innocent lives. For Oppenheimer, this indifference made post–World War II uses of atomic weapons more likely.[9] Oppenheimer's perception was validated again as the United States and Soviet Union conducted a decades-long nuclear arms race.[10] To compensate for their conventional military disadvantage relative to the Soviets, the United States built a large nuclear arsenal, developed nuclear war plans, and leveled nuclear threats at Moscow as a principal means of securing its homeland and allies. The Soviet Union responded in kind. A handful of other countries built smaller but nonetheless powerful nuclear arsenals: the United Kingdom, France, China, South Africa, and Israel. The superpower politics of nuclear brinkmanship brought the world close to nuclear war in 1962 during the Cuban Missile Crisis, when the Soviets placed nuclear-tipped missiles aimed at the United States in Cuba. After a tense standoff, the Soviets removed these missiles and the United States reciprocated by removing their nuclear-tipped missiles from Turkey.[11] A second nuclear reckoning was narrowly avoided again in the 1983 Able Archer Crisis, in which a U.S. nuclear war drill with its NATO allies almost triggered a decision by Soviet leaders to prepare a nuclear response.[12]

A nuclear ethics literature gradually emerged in this Cold War context. In general, academic and policy-oriented contributors to this literature expressed a nearly unified condemnation of unrestrained or "all-out" nuclear warfare— with perhaps the exception of a few Christian apocalyptic writers who could claim neither policy expertise nor widespread respect within the academic community.[13] And once it became evident that the military use of nuclear weapons was increasingly unlikely,[14] the mainstream commentators' moral concern narrowed to issues of nuclear deterrence and the possibilities of lim- ited nuclear reprisal or retaliation.

This chapter reviews the English-language nuclear ethics literature devel- oped during the Cold War on these policy questions.[15] Its evolution reflects an overwhelming concern with U.S. nuclear policy given Washington's oversized role in the international nuclear order.[16] Specifically, many commentators perceived a necessary connection between the moral permissibility (or impermissibility) of U.S. or Western nuclear deterrence *and* retaliatory or defensive nuclear policies. Others rejected the necessity of this connection. Ultimately, this lack of consensus was based on a fundamental disagreement over which moral principles or warrants had privileged status for analysis and assessment. The chapter first considers a moderately diverse set of just war and pro-deterrence consequentialist views based on the rights of states or peoples of survival or the preservation of their way of life. Afterward, it turns to Kantian-inspired anti-deterrence views based on a fundamental respect of the absolute right of noncombatants to life and freedom from the threat of harm. It then examines utilitarian critiques of deterrence based on the consequentialist moral duty to realize the global greater good that, in this case, meant the survival of humanity. Finally, this chapter will show that the Cold War nuclear ethical debates were largely unresolved, and this finding will set the stage for the third chapter's analysis of the post–Cold War era of the nuclear age and the question of the need for a new twenty-first-century nuclear ethics.[17]

Before the next section sets the stage for the chapter's literature review, a proper sensitivity to evolving scholarly norms makes it important to acknowledge that most contributors to the nuclear ethics literature during the Cold War were white men. Most, if not all, feminist approaches to International Relations perceive this racial and gender imbalance among contributors to have contributed significantly to the widespread support in the literature for nuclear deterrence and defense policies.[18] Of course, women and persons of color had long been active in peace and disarmament movements.[19] But, with few exceptions (e.g., Sisela Bok, Jean Bethke Elshtain, Elizabeth Anscombe), the volume and visibility of their contributions to this literature did not capture notice in the academic or policymaking mainstream until the post–Cold War era.

STAGE SETTING: DETERRENCE THEORY

It is important to review briefly some key elements of deterrence theory before analyzing the Cold War–era nuclear ethical debates.[20] The chief purpose of threatening violence is to dissuade a rival or enemy (i.e., a deterree) from undertaking action hostile to the deterrer's perceived national survival and security interests. The dissuasive effort is fundamentally about affecting the deterree's state of mind: that is, to cause the deterree to believe that acting contrary to the deterrer's vital interests will result in unacceptable degrees of punishment or the denial of victory. Indeed, the deterrer prefers a strategy of dissuasion since, if effective, it preserves national security interests without the heavy and perhaps catastrophic costs of armed defense.

It is important to note that the deterrer's calculation of a threat's effectiveness is probabilistic—that is, the deterrer cannot be certain that the deterree will perceive the threat as credible. This distinction between credible and incredible threats suggests a set of necessary (and possibly jointly sufficient) conditions for an effective deterrence policy. One condition is the deterrer's possession of sufficient military capabilities to carry out the threat, such as the capability to deliver such weapons to target. Another is that the deterrer must form the (conditional) intention to carry out the threat if the deterree misbehaves.[21] In other words, it is not rational to threaten the use of nuclear force and not be willing to carry out the threat, unless one intends to implement a bluffing strategy. Finally, threat credibility is closely related to the proportionality of the threatened action to the deterrer's interests. Generally, threats that promise "overkill" responses are less credible, especially if the deterrer's expected costs are high.[22] As will be discussed later, such questions commanded significant attention in the Cold War–era nuclear ethics literature.

To these three necessary conditions, there is one sufficient condition of effective deterrence: that the deterree is convinced by the threat and chooses to refrain from acting against the deterrer's vital interests. The only direct proof of effective deterrence is the deterree's admission to this effect; but it is rare for a deterree to openly admit being deterred and suffer the likely political costs (e.g., loss of status, loss of bargaining leverage in future disputes). Deterrence theorists have thus acknowledged the difficulty of proving why nonevents—such as the absence of aggression—have "occurred."[23] Even so, William Walker's analysis strongly suggests that the superpowers and their allies were committed to the idea that nuclear deterrence worked and was therefore a necessary constitutive feature of the international order of nuclear restraint.[24]

The discussion so far implies that the political justification of nuclear defense and deterrence policies are grounded on the value of state survival and security and, accordingly, their strategic justification is grounded on a rational anticipation of these policies' success. The corresponding question that occupied nuclear ethicists during the Cold War might be framed as follows: if nuclear defense and deterrence are justifiable on political and strategic grounds, are they also justified on moral grounds? Commentators' lack of agreement on this question was rooted in their disputes over which of the competing moral principles carried the status of being non-overridden.[25] The next section reviews the moral arguments that privilege the status of a national survival and security imperative, and the following section reviews those which privilege the imperative to preserve noncombatant lives and rights (i.e., human security). The chapter's conclusion discusses the implications of these unresolved moral debates for nuclear ethical analysis.

THE MORAL IMPERATIVE OF NATIONAL SURVIVAL AND SECURITY: JUST WAR AND PRO-DETERRENCE CONSEQUENTIALISM

Those commentators who privileged the imperative of national survival and security tended to identify with one of two schools of thought: the broad just war tradition and the rationalist tradition linked to Hobbesian political theory. Between these two sources of influence, the just war tradition was—and still seems to be—the primary approach through which moral debate on nuclear weapons policy is conducted.[26] Just war commentators believed a proper moral assessment of nuclear deterrence and defense policies required an explication of the foundations of *jus ad bellum* principles that function to justify their claims on the overriding value of state survival and security. By contrast, the rationalists tended to assume the right of state survival and security as given and did not bother to provide any further justification of those values.

The Moral Foundations of Just War and Rationalist Approaches

For religious commentators in the just war tradition,[27] the *jus ad bellum* principle of the right of national self-defense prescribed the use of armed force as a last resort given the inescapable natural law duty of governments to protect the innocent civilians under their charge.[28] Moreover, the right of resort to armed force entailed the right to threaten its use.[29] As the term "just war" intimates, the defensive use of force is consistent with the demands of domestic and international justice (see chapter 6 on the order-justice dilemma). As the U.S. Catholic Bishops contended in their 1983 *Challenge of Peace*, "Justice is always the foundation of peace. . . . [W]ar is permissible only to confront a real and certain danger, to protect innocent life, and to preserve decent societal conditions and basic human rights."[30] Others, like Elizabeth Anscombe and Keith Payne, emphasized the biblical contention that God had authorized governmental officials to use force to secure peace and justice by punishing wrongdoers.[31] On both accounts, the underlying moral argument begins with an initial condition of injustice constituted by an enemy's aggression. Although lethal military force is *prima facie* evil, it is nonetheless justified, as well as the threat of the same, in the effort to protect innocent lives and decent society from aggression and its harmful effects.

By contrast, the starting points of Walzer's secular just war argument are found in a moral intuition (independent of divine or natural law) about the fundamental rights of individuals and their political communities to survival and security. Uncertain about the epistemological status of "moral foundations," Walzer nonetheless contended,

> Individual rights (to life and liberty) underlie the most important judgments that we make about war. How these rights are themselves founded I cannot try to explain here. It is enough to say that they are somehow entailed by our sense of what it means to be a human being. . . . States' rights are simply their collective form. . . . Over a long period of time, shared experiences and cooperative activity of many different kinds shape a common life. "Contract" is a metaphor for a process of association and mutuality, the ongoing character of which the state claims to protect against external encroachment. . . . Hence, it is a moral process, which justifies some claims to territory and sovereignty and invalidates others.[32]

Walzer's agnosticism about the philosophical foundations of individual and collective self-defense rights might appear to religious commentators as sidestepping a necessary condition of moral justification for the use of defensive force. However, Walzer seems to believe that the only necessary and sufficient condition for justifying the rights of national survival and security is the mutual recognition among members of an international society that their way of life deserves to be protected from unwarranted encroachment. This mutual recognition is formalized in international law, which recognizes the rights of states to territorial integrity and political sovereignty *and* which permits states to use armed force in self-defense.[33] Walzer does not seem to believe that the socially constructed character of this warrant minimizes its justificatory force. Rather, it seems to synthesize a time-honored and respected tradition of justification for the resort to (the threat of) armed force with more contemporary philosophical sensibilities.

In contrast to the religious just war commentators, Rationalists conceive the foundations of morality differently. Rather than emerging from divine command as special revelation or couched within natural law, they understood the foundations of moral action as intimately linked to the realization of subjective goods. The individual goods of survival and security are taken as logically prior values from which all other values are derived. For instance, the related values of collective survival and security are a function of the aggregation of these individually held goods. When assessing actions relative to the achievement of goods or benefits, David Gauthier proposes,

> Morality . . . follows rationality. Practical rationality is concerned with the maximization of benefit; the primary requirements of morality are that, in maximizing benefit, advantage must not be taken and need not be given. . . . [As Hobbes

enjoined, it is prudent to] "seek peace and follow it" and "by all means we can, to defend our selves." Hobbes understands that these requirements are mutually supportive.[34]

Hence, the only requirement for morality as rationality to justify a decision to use force, or threaten it, is to show the corresponding maximization of these benefits.[35] Furthermore, Gauthier's emphasis on self-defense "by all means we can" suggests a line of justification for certain courses of action, like the use of nuclear reprisal strikes if deterrence fails, which challenges the humanitarian elements of just war theory.

On the Moral Defense of Nuclear Deterrence: Points of Consensus and Dissensus

Just war and rationalist commentators asserted the moral permissibility or necessity of nuclear deterrence to prevent the outbreak of great power (nuclear) war. Beyond this point of agreement, they differed on the moral appropriateness of targeting nuclear weapons against an adversary's cities (countervalue deterrence) as opposed to their military assets (counterforce deterrence). They also differed on the moral appropriateness of nuclear reprisal strikes in case of deterrence failure. These disagreements were based on competing understandings of the authority or force of moral side constraints on military force, which in the just war tradition are identified as *jus in bello* principles. One is the noncombatant immunity or discrimination principle, which proscribes the deliberate targeting or use of force against innocent civilians.[36] Another is the proportionality principle, which proscribes the use of excessive force to attain a military objective.[37] These disagreements persisted despite the fact that the *jus in bello* principles eventually were codified in positive international law.[38] The next several paragraphs examine these points of consensus and dissensus.

Points of Consensus

Given the right of national self-defense, the focal political issue that engaged Western nuclear ethicists was the prospect of a Soviet invasion that the United States and its allies might not be able to repel and defeat with conventional force alone.[39] For Sir Michael Quinlan, whom Frank Jones called the "high priest of deterrence" in the just war tradition, the goal of Western nuclear policy was to "prevent any war, not just nuclear war, between East and West."[40] Similarly, for Keith Payne the goals of U.S. nuclear deterrence were to prevent the Soviet Union from (1) highly provocative or coercive behavior that might erupt into war, (2) nuclear or conventional attacks on U.S. allies or friends around the world, and (3) nuclear or conventional attacks on the U.S. homeland.[41] Gauthier's rationalist account also assumed the necessity of nuclear deterrence based on anticipated Soviet preferences to gain territorial advantages in Western Europe if it could be done without suffering significant retaliation.[42]

The strategic discourse that justified nuclear deterrence sometimes masked a fear of losing or surrendering the West's "way of life" to Soviet conquest. For just war commentators, this fear was linked to an apprehension over a prospective failure to preserve freedom of religion as one of the foremost democratic liberties, even more so than preserving the West's territorial integrity.[43] The focus on securing a way of life has been called the pursuit of "ontological security," which is distinct from the security conceptions associated with "territorial integrity" and "sovereignty."[44] As chapter 4 discusses, the commitment to ontological security as an non-overridden moral value clashes with the commitments of deterrence critics, who take universal human rights (or human security) as a non-overridden moral value.

A second point of consensus focused on nuclear deterrence as a morally salient risk-reward strategy. In Joseph Nye's view, national survival and security have logical priority when it comes to defense policy, but they are not absolute values.[45] He mentions that, paradoxically, the values that make life worth living (e.g., democratic practices, cherished freedoms of thought and expression) sometimes conflict with a desire for absolute safety from harm. Thus, for Nye the appropriate question when considering the morality of nuclear deterrence is if it is "possible to imagine any threat to our civilization and values that would justify raising the threat to a billion lives from one in ten thousand to one in a thousand for a specific period?"[46] In his affirmative answer, Nye did not mean for the comparison between 1/10,000 and 1/1,000 to be understood as precise risk assessments; even so he believed that the odds of nuclear war were significantly lower than alarmists believed.[47] Indeed, Nye emphasized that nuclear crises had not escalated into catastrophes but rather produced nuclear restraint:

> Human interactions are more like loaded dice. The odds change, and the outcome of one set of events may greatly change the odds for the next event. In fact, frightening events like the Berlin or Cuban missile crises may drive the odds of war down in their immediate aftermath. . . . The likelihood of nuclear war rests on both independent and interdependent probabilities that relate to different aspects of the process by which war might occur.[48]

By citing the Berlin and Cuban missile crises between 1958 and 1962 as defining moments of nuclear brinkmanship, Nye suggested that future nuclear crises could likely lead to the further entrenchment of nuclear restraint.[49] And even though the 1980s witnessed an initial but notable increase in nuclear anxiety among many peoples as President Reagan abandoned détente for anti-Soviet stridency and increased nuclear spending, Nye maintained that the odds of nuclear war remained quite low and therefore morally permissible.[50]

Even so, one of Nye's key maxims of nuclear ethics emphasized the importance of working to "reduce the risks of nuclear war in the near term" even if they could not be eliminated altogether.[51] The policy principles that would guide such reductions included expanding conventional deterrence to prevent lower-level conflicts (which nuclear deterrence could not credibly address), more effective crisis management (so that tense standoffs between adversaries would not devolve into any one state feeling desperate and therefore willing to use nuclear weapons), and vigorous arms control and nonproliferation agreements to prevent irresponsible state actors from acquiring nuclear weapons.

Gauthier concurred by discussing if a deterrer could correctly anticipate a deterree's response to nuclear threats.[52] Gauthier acknowledged that, in deterrence failure, carrying out a nuclear threat would most likely not be "utility-maximizing," especially if it led to escalatory responses. Even so, Gauthier defended the rationality (i.e., morality) of nuclear deterrence by arguing the following:

1. It may be utility maximizing to form the retaliatory intention; therefore
2. It may be rational to form such an intention; therefore,
3. If it is rational to form the intention, it is rational to act on it; therefore,
4. A rational actor can sincerely express the intention; and therefore
5. Another rational adversary can be deterred by such an expression.[53]

Gauthier's argument assumes the adequacy of the deterrer's capabilities and focuses on the causal link between his intention to retaliate and the deterree's choice to avoid aggression. Since the deterrer's knowledge of the deterree's responses to nuclear threats is probabilistic (and therefore risky), Gauthier argues that the formation of this intention leaves the deterrer no other rational (or moral) option than to "stick to their guns" if deterrence fails.[54] It follows that the deterrer's commitment to carry out a nonutility maximizing act could cause the deterree to believe the nuclear threat, choose to avoid aggression, and ultimately to show that deterrence was in fact utility maximizing.

The choice to "stick to one's guns," though, carries its own moral risks and paradoxes that another rationalist, Greg Kavka, identifies in his account of the paradoxes of nuclear deterrence. The first paradox follows Gauthier's account in that it would be right to form an intention to act in grossly immoral ways if it was required to make deterrence succeed.[55] Ordinarily, it is wrong to intend to do something that is wrong to do; however, Kavka contends that

in certain cases, intentions may have *autonomous effects* that are independent of the intended acts actually being performed. In particular, intentions to act may influence the conduct of other agents. When an intention has important autonomous effects, these effects must be incorporated into any adequate moral analysis of it.[56]

Kavka's contention here raises a few skeptical questions important for his pro-deterrence argument. First, one might ask if intentions that are not enacted can still have causal effects on a deterree's choices and actions. Is it possible for a nonaction to produce effects? Or, is it more plausible to believe that the deterree, having perceived that the threat is credible, is solely responsible for the movement away from hostile action against the deterrer?[57]

These skeptical questions might be satisfactorily answered if we read Kavka's contention as implicitly referring to "expressed intentions." In general, speech act and securitization theories argue that discourse has "perlocutionary" effects: that is, the act of speaking can cause changes in social conditions, such as the behavior of other actors.[58] Thus, we can accept a reading of Kavka's contention which states that the deterrer's expressed intentions to engage in nuclear reprisal is a speech act that can have autonomous effects on the deterree's action independent of any implementation of the threat. Accordingly, for Kavka, the deterrer's threat expresses an intention that can cause the deterree to change course of action and thus eliminate the need to carry out the nuclear threat.

The second of Kavka's paradoxes of nuclear deterrence is that a "rational and morally good agent cannot (as a matter of logic) have (or form) the intention to apply the sanction if the offense is committed."[59] For Kavka, an action is "right" if and only if a morally good person would choose that same course of action in a given situation.[60] In the case of nuclear deterrence, though, the kind of action that is right as defined by the first paradox is also that which is abhorrent to a morally virtuous actor. Taken together, a deterrer who is morally virtuous will then encounter a grave moral dilemma—their duty to act rightly will necessarily involve their own moral corruption.

Kavka's third paradox of nuclear deterrence is that "it would be morally right for a rational and morally good agent to deliberately attempt to corrupt himself" by (attempting to) form the intention to inflict nuclear retaliation on the deterree, if necessary.[61] If the deterrer is truly rational and morally good, the process of self-corruption would be quite difficult; even so, the deterrer presumably would know that it would be morally wrong to avoid self-corruption or seek moral rehabilitation afterward. Any agent that prefers their own moral purity over the duty to prevent a nuclear calamity must be regarded as selfish and morally stunted. Thus, an agent (i.e., commander-in-chief) who could not trust himself to "stick to his guns" might instead delegate launch authority to subordinate officers or to a reliable automated process.

The upshot of these arguments on the moral justification of nuclear deterrence as a risk strategy is that the just causes of national security and war prevention are worth the cost of moral purity, all things considered. Given the realistic but probabilistic autonomous effects of expressed intentions of nuclear retaliation on deterree behavior, the commentators agreed that the commission of such wrongful speech acts paradoxically had redounded to the moral good.

Points of Dissensus

Notwithstanding these points of agreement, just war and rationalist commentators disagreed profoundly on the proper limits of nuclear deterrence and defense. The focus of their disagreements was on the proper application of the *jus in bello* principle of discrimination—that is, the degree to which its constraints must be relaxed in cases of national emergency.

On a strict understanding of the discrimination principle, the scope of nuclear threats ought to be limited to a deterree's military assets (i.e., counterforce deterrence).[62] For Elizabeth Anscombe and Paul Ramsey, this was the only morally justified policy of deterrence, since counter-city or countervalue deterrence was unconscionable.[63] Ramsey advanced two arguments by analogy in support of this contention that captured significant attention among other commentators. He imagined a case where it could be proven that tying infant human beings to the front bumpers of automobiles would effectively prevent the deaths and injuries caused by automobile crashes. Such policy would nonetheless express a blatant disregard for the rights of innocent life, leading Ramsey to argue it could not be morally justifiable.[64] On the same grounds, Ramsey argued against any practice resembling armed and feuding families who target each other's children to prevent future clan violence (e.g., the Hatfields and McCoys).[65] If it is never right to keep anyone's children in a sniper's crosshairs, it is also not right to keep Russian (or any other country's) children as nuclear hostages. Keith Payne echoed Ramsey's view by asserting that any Christian embrace of counter-city nuclear deterrence was altogether inconsistent with *jus in bello* obligations. Instead, Payne contended that discrimination in targeting could send a morally sound and strategically prudent signal of willingness to act with restraint in a crisis, thus reducing the reciprocal risk of harm to innocent Americans held hostage by Soviet missiles.[66]

In contrast, looser readings of the discrimination principle suggested that counter-city deterrence was morally permissible or excusable.[67] Two such readings can be distinguished: "threaten but do not use" and "threaten and use if necessary." An example of the former reading is Fr. J. Bryan Hehir's view that "deterrence prohibition" did not follow from the discrimination principle once we "acknowledge[d] the empirical realities of strategic policy."[68] Indeed, these realities were that the United States and Soviet Union employed both counter-city and counterforce deterrent strategies (as well as war plans to "win" nuclear wars) during the most of the Cold War.[69] Hence, on Hehir's view the ends of deterrence policy (war prevention, protection of the innocent) were different from those of strategic policy (win wars or limit the damage suffered by them), which in turn suggested a difference in the application of the discrimination principle. Thus, while strategic policy commits state officials to the use of nuclear weapons, deterrence policy does not. If deterrence policy is successful, then the resulting avoidance of nuclear retaliation means the avoidance of nuclear harm on innocent lives. As Walzer had put it, nuclear threats by themselves do not cause injuries.[70]

Those committed to Ramsey's stricter reading of the discrimination principle might wonder if nuclear threats that are not "meant" to be carried out could be expressed without somehow intending to use nuclear weapons if deterrence failed. That is, how could one simultaneously honor the fundamental moral imperative to form intentions consistent with moral virtue and also form the intention to threaten mass annihilation on an enemy's population? To that question, Hehir claimed that the only intention that deterrence policy requires of state officials is to do what is necessary to persuade the deterree to refrain from aggression. When pressed whether his response amounted to endorsing a policy of nuclear bluffing, Hehir seemed to affirm that "the fact of the mere possession of a devastating nuclear deterrence means that the adversary can never be sure it will not be used. In any rational calculation, that should be enough to deter."[71] Hehir's attempt at clarification suggests that the entire force of nuclear deterrence rested on its material capability without any genuine intention to follow through on the implied threat.

It seems right to conclude that Hehir's argument is an attempt to reconcile counter-city deterrence and *jus in bello* requirements. This attempt also became the basis of the U.S. Catholic Bishops' conclusions in their 1983 *Challenge of Peace* that "under no circumstances" should nuclear weapons ever be used "for the purpose of destroying population centers or other predominantly civilian targets" and that

> in current conditions "deterrence" based on balance, certainly not as an end in itself but as a step on the way toward a progressive disarmament, may still be judged morally acceptable. Nonetheless, in order to ensure peace, it is indispensable not to be satisfied with this minimum which is always susceptible to the real danger of explosion.[72]

Taken together, the two statements enjoin the United States and other nuclear-armed powers to refrain from ever using nuclear weapons. However, they also permit the United States (and presumably other nuclear powers) to express the threat of nuclear use as an interim measure on the road to nuclear abolition. For, as they warn, the "real dangers" of catastrophic nuclear war can never disappear as long as the superpowers and some of their allies retain nuclear weapons.

By contrast, Quinlan and others asserted that "threaten but do not use" was an incoherent national security posture, and that the only morally viable nuclear deterrence strategy was to "threaten and use if necessary." For Quinlan, "deterrence cannot exist if there is no possibility of use," and for this reason Hehir and the U.S. Catholic Bishops' "stance can scarcely be accepted as solving a profound moral dilemma."[73] For his part, Nye's rejection of "threaten but do not use" focused on institutional requirements. Since nuclear deterrence requires "a complex bureaucratic machinery," an effective nuclear bluff would be impossible to conceal.[74] For Nye, it is unlikely that a president or prime minister could keep a nuclear bluff secret while everyone else remained convinced that their nuclear threats were to be implemented if deterrence failed. Indeed, he believed it would be more likely that the commander-in-chief's intentions would become known by accident or design, leading to the very opposite of what deterrence policy was meant to produce.[75]

In hindsight, it is obvious that the debate over nuclear deterrence intentions bore most prominently on whether the discrimination principle was best understood strictly or more loosely. In just war theory, this debate arises in cases where the looser application of *jus in bello* constraints seems to undermine a war justified by *jus ad bellum* rights or requirements. To address this problem, the doctrine of double effect functions as an *ad hoc* principle to weaken or relax the *jus in bello* principles. Thus, it permits or excuses armed force harmful to innocents provided (1) it would have been morally permissible otherwise, (2) the forceful actor intends to do good and not evil, (3) the doing of evil is not a means to the doing of good, and (4) the goodness derived from undertaking military action outweighs the evil done to the civilian population.[76] If nuclear deterrent threats are intended for good (the prevention of war, the prevention of Soviet conquest), the salient and related question was: "how could it ever be consistent with the imperatives of discrimination and proportionality to use the appalling destructive power of nuclear weapons [to restore deterrence in case of deterrence failure] in ways that would be certain to inflict heavy non-combatant deaths?"[77]

Quinlan's response to this question gestured toward what was earlier called "ontological security" by comparing the evils of massive noncombatant deaths in adversary states to the evils of Westerners living and suffering under Soviet rule. In doing so, Quinlan conceded that using nuclear weapons was likely to be grossly indiscriminate but nonetheless proportionate in relation to the value of preserving Western society:

> It is unclear that the scale of killing in strategic nuclear action would be disproportionate to any good result that could reasonably be expected. . . . Deaths on a huge scale would be an appalling calamity. But, proportionality involves not one

factor only, but the relationship between two. World conquest or domination—even for a short time, as of the Nazi conquests—by a tyranny like those of Hitler and Stalin would also be an appalling calamity.[78]

For Quinlan, the dual calamities of "deaths on a huge scale" and subjection to a "tyranny like those of Hitler or Stalin" "even for a short time" must be weighed against each other to determine if nuclear deterrence and nuclear reprisal were incontrovertible violations of the principle of proportionality. Interestingly, not only did Quinlan affirm the "threaten but use if necessary" reading of the discrimination principle, but he seemed to subsume the proportionality principle to the demands of ontological security.

Quinlan's use of the doctrine of double effect and the proportionality principle to justify the nuclear defense of the West against totalitarian tyranny did not win universal Catholic agreement. For her part, Elizabeth Anscombe condemned such approaches as a misuse of the doctrine of double effect and a (deliberate) misunderstanding of Scripture:

> Some Catholics are not scrupling to say that anything is justified in defence of the continued existence and liberty of the Church in the West. A terrible fear of communism drives people to say this sort of thing. . . . Those, therefore, who think they must be prepared to wage a war with Russia involving the deliberate massacre of cities, must be prepared to say to God: "We had to break your law, lest your Church fail. We could not obey your commandments, for we did not believe your promises."[79]

On Anscombe's view, approaches such as Quinlan's overlook that the Christian Scriptures do not endorse an "ends justify the means" approach regarding the defense of the Church in the West and the wider defense of Western society. The divine commandment against murder was absolute, and the obliteration of Soviet cities in the defense of the Western Church must be counted as murder. If it were to occur, the Church's demise was not to be found, as Quinlan seem to have believed, in its suppression by a future Soviet tyranny. Rather, it was found in the attitude and actions of Christian combatants who would have to confess that they broke God's law in the name of preventing the Church from failing. Thus, for Anscombe, the prospect of Soviet domination was not to be feared more than "the destruction of people's bodies by obliteration bombing."[80]

Walzer's approach to the discrimination principle is located somewhere in between Anscombe's and Quinlan's, although his ultimate observation cited in the chapter's introduction that nuclear weapons explode just war theory foreshadows the unavoidable nihilism of nuclear warfare. Walzer accepts the wrongful intentions principle (WIP) that, in ordinary circumstances, it is wrong to threaten to do what is clearly wrong to do. For instance, Walzer acknowledged it would be wrong for local police to form the intention to kill the family members and friends of murderers to deter any future murders, and then to express that intention in the form of a threat.[81] However, he believed the prospect of a "supreme emergency" can change the conditions under which WIP is applied. A supreme emergency is that condition where a state faces an imminent existential defeat which it cannot repel except by means that are ordinarily impermissible.[82] For Walzer, the Cold War realities of superpower policies of mutually assured destruction have made the supreme emergency a "permanent condition."

> Deterrence is a way of coping with that condition, and though it is a bad way, there may well be no other that is practical in a world of sovereign and suspicious states. We threaten evil in order not to do it, and the doing of it would be so terrible that the threat seems in comparison to be morally defensible.[83]

With these words, Walzer's moral defense of nuclear deterrence is less confident than Quinlan's. Walzer seems to say that it is not that nuclear deterrence *is* morally justifiable; it is that it *seems so* in comparison to the horrific prospect of carrying out the nuclear threat. Indeed, he seems to depart from Quinlan significantly on the question of nuclear reprisal and finally conclude that it cannot be reconciled with morality:

> Nuclear weapons explode the theory of just war. . . . [O]ur familiar notions about *jus in bello* require us to condemn even the threat to use them. And yet there are other notions, also familiar, having to do with aggression and the right of self-defense, that seem to require exactly that threat. So we move uneasily beyond the limits of justice for the sake of justice (and of peace).[84]

In the end, Walzer's argument is that nuclear weapons have subverted morality and perhaps even the justice of national self-defense, even if it could be shown that nuclear deterrence was necessary to avert the annihilation of humanity during the Cold War era. Unsurprisingly, Quinlan and Walzer agreed that it was morally necessary for the superpowers to find a diplomatically administered process for ridding the world of nuclear threats peacefully.[85] Until then, however, any embrace of nuclear deterrence would be morally paradoxical and deeply problematic.

THE MORAL IMPERATIVE OF HUMAN RIGHTS: KANTIAN APPROACHES

In contrast to the just war and rationalist arguments reviewed in the previous section, Kantian-inspired commentators argued that the moral priority of human rights overrode or at least were equally forceful against any competing imperative, such as national survival or security. On this foundation, these commentators condemned nuclear defense and deterrence in absolute or near-absolute terms. In so doing, they revealed some important points of convergence on the absolute nature of noncombatant immunity with just war commentators like Ramsey and Payne. Even so, their cosmopolitan commitments to human rights clearly distinguished Kantian-inspired commentators from their just war interlocutors' preference for more exclusive national security commitments.

In the next few paragraphs, the moral foundations of the Kantian approaches to nuclear ethics are reviewed and then applied to the Cold War policies of nuclear deterrence and reprisal. Special emphasis is given to arguments concerning the limits on the self-defense argument, the murderous nature of nuclear threats, and the immorality of nuclear hostage-holding.

The Moral Foundations of Kantian Approaches

For Kantian commentators, the categorical imperative is the starting point for any moral analysis of nuclear deterrence and reprisal policy. In Kantian moral theory, the categorical imperative is the supreme law of moral reason, and it provides a universal and timeless method of practical reasoning. Kant contrasted the categorical imperative with "hypothetical imperative" whose prescriptions for action were conditional and therefore were not considered absolutely binding. For Kant, the supreme law of moral reason could not be relativized, amended, or excused; otherwise, it would be impossible to require actors to adhere to moral duty when it conflicted with self-interest. Additionally, Kant contrasted the categorical imperative with divine commands by emphasizing the role of the autonomous human actor in self-prescribing rules or maxims for action consistent with the categorical imperative. In this way, Kant affirmed the value of human freedom while also recommending a secular equivalent of the Gospel commandment "seek ye first the kingdom of God and his righteousness, and all else [e.g., national survival and security] will be added to you."[86] Thus, the application of the categorical imperative by each autonomous person in private life and political society would preserve the moral freedom of all persons who could mutually affect one another, leading to the realization of individual and collective security within political society. Indeed, according to Sisela Bok, Kant's in/famous motto "do what is right though the world should perish," often ridiculed by consequentialists as naïve utopianism, might actually reveal an important insight that is key to the premise of this book related in chapter 1 on the relationship of justice to security: "If [a given action] is right [or just] it cannot precipitate the end of the world."[87]

Kant expressed three formulations of the categorical imperative to address both ideal and nonideal conditions for moral theory.[88] Two of these formulations were emphasized by Cold War–era commentators: the Formula of Universal Law and the Formula of Humanity.[89] The Formula of Universal Law required an agent to "act only in accordance with that maxim through which you can at the same time will that it become a universal law."[90] On this formulation, an agent constructs a maxim for their own action such that they could accept its constraints on their own self-interest (and not only others' self-interest) if it were to be codified as a universal law. For Kant, the universality condition of the categorical imperative was necessary to ensure that one's proposed maxims were not inconsistent by simultaneously requiring and forbidding the same course of action for any similarly situated actor.

Alternatively, the Formula of Humanity required an agent to "act in such a way that you treat humanity, whether in your own person or in the person of another, always at the same time as an end and never simply as a means."[91] Although this formulation was logically consistent with the Formula of Universal Law, its substance related a stricter moral obligation. The focus of the Formula of Humanity was on the dutiful treatment of human persons always and in all places as ends in themselves, even if some relationships also required treating persons as means to one or more individual or collective ends. Thus, in institutional contexts where an organization's mission obliged subordinates to obey the orders of superiors, these orders were not to exceed the organization's mandate by prohibiting subordinates from private or public action independent of the organization. For instance, superiors wronged subordinates by preventing the latter's free expression of belief.[92]

Applied to questions of national security, the Formula of Humanity prohibits absolutely the mistreatment of innocent persons during wartime or conflicts short of war. It follows from this formula that nuclear war is absolutely prohibited. Furthermore, a similar application of the Formula of Humanity seems to absolutely prohibit nuclear threats. Interestingly, some Kantian commentators entertained the possibility of a conditional condemnation of nuclear deterrence. This nuance among some Kantian accounts reflected a significant sympathy with Walzer's depiction of the moral dilemmas of a supreme emergency condition.[93] In that regard, we can distinguish between three lines of Kantian argument on this issue: the immorality of unconstrained national self-defense, the immorality of murder and murderous threats, and the immorality of nuclear- hostage-holding.

The Immorality of Unconstrained National Self-Defense

Although just war commentators such as Quinlan did not explicitly refer to the *jus ad bellum* self-defense right as an unconditional or absolute value, it was implicit in their emphasis on the critical need to preserve Western democracy from illiberal aggression. Thomas Donaldson challenged this implicit stance by arguing that an important distinction must be made between genuinely existential threats to state survival, which are rare, and the ordinary cases of defending national interests.[94] Donaldson's distinction reveals an appreciation of the force of Walzer's supreme emergency condition, although it also stresses that the self-defense argument can often masquerade as the self-preservation argument in most justifications of nuclear war and deterrence policy.

In Donaldson's view, the exercise of national self-defense is constrained by the noncombatant immunity and proportionality principles and therefore it prohibits altogether measures such as nuclear reprisal strikes or the threat of them. Against those who have claimed that discriminate and proportional use of nuclear weapons is possible, Donaldson argued that the "critical moral feature of nuclear weapons systems" is "technological recalcitrance."[95] This is to say, the design and explosive capabilities of nuclear weapons make it extremely unlikely that their uses can be discriminate except in rare cases, such as nuclear attacks against submarines in the deep sea or attacks against enemy satellites orbiting the earth.[96] Rather, nuclear strikes against many, if not most, military targets will ultimately involve the deaths of innocent people on "a huge scale."[97] On this basis, to say nuclear reprisal can be consistent with *jus in bello* requirements is clearly false.

Some commentators criticized Donaldson's argument as exhibiting a dubious technological determinism.[98] As Nye saw it, reasonable skepticism that nuclear war could be controlled or managed effectively did not necessarily imply that nuclear escalation was inevitable. Rather, for Nye it meant that governments were still capable of restrained nuclear use, which could well result in fewer casualties than some conventional military operations might produce. Nye maintained that, in such scenarios, limited nuclear defense was thinkable and perhaps morally necessary to preserve one's state from existential defeat.

Interestingly, the Donaldson article that Nye examined suggests a response to the latter's objection. Donaldson recalled the domestic analogy of home self-defense to identify the proper limits on the right of security and to also highlight the problem of technological recalcitrance. He imagined two cases of defense against home invasion by the owner (1) threatening to use a special gun that targets an assailant's chromosomal structure but that also affects others with the same structure and, alternately, (2) posting signs notifying passersby that his property's boundaries are wired to detonate a number of high-explosive charges if trespass occurs—although there was some chance that an earthquake might detonate the devices as well. For Donaldson, the first case represented indiscriminate targeting effects of weapons whose features promised greater precision in destructive capabilities, and the second case represented weapons that were not only indiscriminate but whose risks of suffering accidents were unacceptably high.[99] In neither case would it be morally justifiable to employ those extreme home defense methods since others are available that comport (at least more closely) with the discrimination principle. But, even if reliance on such methods enhanced the effects of nuclear deterrence policy, they should be rejected because "some actions ought not to be done; some risks ought not be assumed; and they ought not to be done or assumed as a matter of principle."[100]

Donaldson's argument against nuclear defense and deterrence policies can be interpreted as a twofold Kantian amplification of the deontological authority of *jus in bello* principles. First, the discrimination and proportionality principles comport with the Formula of Universal Law, since any state would demand for itself and its citizens immunity from disproportionate force and unjust harm. And last, these principles also comport with the Formula of Humanity, since any state would demand for its officials and citizenry the right to be treated as ends and not as means only.

The Immorality of Murder and Murderous Threats

A different line of argument consistent with the Formula of Humanity was expressed by the philosopher Michael Dummett, who regarded the just war principle of noncombatant immunity as a corollary of the absolute moral and legal prohibitions on murder.[101] On this point, Dummett was in complete agreement with Anscombe's strict interpretation of the discrimination principle. Dummett argued that to

> recognize that nuclear warfare is unconditionally wrong, we need to know only two things: that the same moral principles that govern the lives of all of us apply to governments and to what is done at the command of governments; and that moral principles are universal. If a moral principle is valid at all, it is valid for everyone, in all places, in all circumstances and at all times; war cannot suspend moral principles, though it provokes their violation. If the obliteration of whole cities, or whole populations, is not murder, there is no such thing as murder; if it is not wrong, then nothing is wrong.[102]

Dummett's claim that the "same moral principles" apply to individuals and governments alike is consistent with Kant's Formula of Universal Law, and his principled objection to mass murder recalls Kant's Formula of Humanity. Furthermore, his inference that morality cannot identify any act as "wrong" if entire populations can be destroyed by nuclear means takes Walzer's judgment that "nuclear weapons explode Just War theory" to its logical conclusion. For Dummett, not only do nuclear weapons explode just war theory as a system of moral thought, but they obliterate morality altogether. Thus, for

Dummett, "no one committed to defence by nuclear weapons can have any principled objection to murder."[103]

Nye strongly objected to Dummett's argument, classifying it as an instance of shrill and irresponsible moralizing. On Nye's view, advocates and practitioners of nuclear defense did not intend to commit future mass murder; on the contrary, they intended to prevent such an occurrence. Nye urged moral critics to cultivate a greater sensitivity to the strategic contexts that might require nuclear defense rather than making up "a theory about their opponent's motives." Moreover, he urged moral critics to "pay more heed to the strategists' arguments and to realize that they will need to work with more realistic assumptions if they wish to be effective in a dialogue between ethics and strategy."[104]

Nye's point is well taken about the need for greater empathy and sensitivity to points of argument about the morality of nuclear weapons across the academic and policy divide. Unfortunately, his objection to the tone of Dummett's remarks overlooks its substance and logical force. A reconstruction of Dummett's argument can clarify this last point:

1. Our considered moral judgments acknowledge that unjust or undeserved killing counts as murder.
2. If moral principles, and specifically the *jus in bello* noncombatant immunity principle, prohibit murder absolutely and universally, then there is an absolute and universal principled objection to the use of nuclear weapons for national defense, which results in large-scale deaths of noncombatants.
3. However, no one committed to the use of nuclear weapons for national defense acknowledges that there is an *absolute* and *universal* principle prohibiting noncombatant deaths arising from military defense.
   Therefore,
4. No one committed to the defense by nuclear weapons can have a principled objection to murder.

Dummett associates the noncombatant immunity principle with the absolute prohibition on murder and correctly concludes that the killing of innocent civilians is undeserved and therefore unjust. There is no other term than "murder" that rightly describes the killing of innocents by way of nuclear force.[105] Recalling Quinlan and Walzer's acknowledgment that nuclear use could not avoid massive civilian deaths, their choice of the term "deaths" nonetheless elides the concept of justice embedded in this *jus in bello* prohibition. Indeed, Dummett's use of "murder" is not fundamentally about tone or "snide attributions" looking to "stifle dialogue."[106] Rather, it is about achieving clarity and precision in the understanding of the discrimination principle's conceptual entailments.[107]

As Dummett turned to assess nuclear deterrence, he contended that it is morally permissible only if the deterrer knows with certainty that his threat will cause the deterree to abandon a previously chosen policy of aggression.[108] Dummett's judgment on this point recalls Kavka's view on the possibility of autonomous effects of strategic intentions underneath expressed nuclear threats. However, the conditions of international politics rarely, if ever, provide state actors the opportunity to possess certain knowledge of an enemy's future (re)action.[109] Even the knowledge that nuclear threats have worked in the past is not enough to provide certain knowledge of their future effects. In such epistemic conditions, Dummett believed it is practically impossible to justify nuclear deterrence since "there is no circumstance in which nuclear reprisal which leads to the mass murder of millions is morally acceptable."[110]

The Immorality of Nuclear Hostage-Holding

In contrast to Donaldson's focus on the limits of the self-defense and Dummett's focus on the murderous nature of nuclear war and deterrence, Steven P. Lee focused on how nuclear deterrence policies necessarily lead deterrer states to systematically engage in the wrongful treatment of innocent citizens. Although Lee granted that a responsible moral assessment of nuclear deterrence required the inclusion of both consequentialist and deontological considerations,[111] he maintained that "anything less than a large amount of social benefit is not enough to override the moral objection to [an] action on the basis of its being a violation of nonconsequentialist rules."[112]

This statement frames Lee's central principle of the "morality of social institutions" (PMSI), where nuclear deterrence counts as a social institution whose chief aims are self-defense and war prevention. PMSI states, "Social institutions are morally justified only if they achieve their social benefit in a way that does not systematically violate nonconsequentialist rules, such as those of justice and the respect for rights."[113] On this principle, Lee assumes that one central duty of every government is to protect and preserve the civil and human rights of their citizens. Lee also distinguishes between policies of direct threats against aggressor governments and threats against an aggressor's citizenry as a way of coercively inducing that government to change its policies. For Lee, the latter kind of deterrence commits the deterrer to a systematic policy of vicarious punishment for an aggressor's wrongdoing. On this distinction, it is not wrong to intend to punish an aggressor or deny the aggressor a victory—and for this reason Lee contends that the intentions behind counterforce nuclear deterrence do not *prima facie* violate the WIP. However, counter-city nuclear deterrence, even if it is successful at preventing aggression, is an instance of "making a third-party threat" which would be wrong to (intend to) do. Thus, for Lee a policy that uses third-party nuclear threats to coerce an enemy state away from an unwanted action cannot avoid being "an institution of hostage-holding."[114]

Lee's account of the immorality of nuclear hostage-holding is consistent with the Kantian Formula of Humanity. Instead of nuclear-armed states treating innocent civilians of an enemy state as ends in themselves, nuclear deterrence policy treats them only as a means of the deterrer's national security interests. Accordingly, his account recommends that nuclear deterrence violates the Humanity Formula in three ways.[115] One is that innocents who are threatened are not responsible for their government's hostile policy against the deterrer. Hence, the use of innocent third parties as levers of control on the deterree's government is purely instrumental and without due regard for the innocents' rights. This is the case even if counter-city nuclear deterrence effectively prevents aggression:

> The moral wrongness of hostage-holding results from the fact that . . . it imposes a risk of harm on the hostages [without their consent], whether or not the potential harm is actualized through the threats being carried out. Otherwise, the moral wrongness of hostage holding would be dissolved by the success of the threat, which is absurd.[116]

With this remark, Lee argues that innocent third parties who have become the object of threats of vicarious punishment cannot choose to escape the additional risk of harm to their own lives imposed by the deterrer. And while nuclear threats may not cause nuclear hostages any physical injuries or otherwise constrain their freedom of movement, they suffer the injustice of an increased risk of harm that they otherwise would not bear. Lee acknowledges that an innocent party might voluntarily accept greater risk to their lives for the benefit of other persons; but governments do not solicit the consent of foreign enemy citizens to accept increased risks against their lives in times of war nor, we might add, do they solicit the consent of their own citizens to become hostages of an adversary state in cases of mutual nuclear deterrence. Thus, there is nothing about the institutional practice of nuclear deterrence that could be considered morally permissible.

It might be objected that PMSI is unreasonably strict if, as Quinlan maintained, it is necessary to prevent two or more hostile great powers from engaging in nuclear war. On this objection, counter-city nuclear threats might provide a social benefit of such magnitude that Kantian concerns should be suspended about the undue imposition of risk against innocent civilians. Lee offered an amendment to PMSI in anticipation of this objection. It states that an institution that systematically violates nonconsequentialist rules is nonetheless morally justified only if (1) this institution achieves a sufficiently great social benefit that (2) could not otherwise be achieved without systematic violation of nonconsequentialist rules and for which (3) there are no other alternative institutions to secure this benefit whose violations of nonconsequentialist rules are less severe.[117] This amended PMSI (hereafter, PMSI*) is consistent with the Kantian Formula of Universal Law insofar as it tolerates a nonideal moral condition (i.e., the systematic violation of the Formula of Humanity) if there is no other plausible means to achieve a "sufficiently great social benefit." The survival and security of humanity would constitute such a sufficiently great social benefit, and thus for Lee the test of nuclear deterrence as it is practiced depends on PMSI*'s second and third criteria. And, for the United States and Soviet Union, this meant testing the practices of national and extended nuclear deterrence policy.[118]

Lee considered first the practice of extended nuclear deterrence, a collective security measure employed by the United States that promised its key Cold War allies to defend their territories from Soviet aggression by nuclear means. Lee claimed that, on the third criteria of PMSI*, extended nuclear deterrence could not be morally justified since the prevention of Soviet aggression might also be achieved by an enhanced and allied conventional deterrent.[119] Secondarily, Lee argued that counterforce nuclear deterrence violated PMSI* despite its promise of significantly greater adherence to the discrimination principle.[120] Counterforce nuclear threats introduced instability into the deterrer–deterree relationship such that the latter might well believe such threats were not defensive in intent but preventive—that is, signaling a willingness to undertake a decapitation strike against the deterree's leadership. The deterree's corresponding security dilemma might lead them to increase their nuclear capabilities, leading to an arms race and an increased risk of general nuclear war. The policy of counterforce nuclear deterrence would thus undermine itself.

In the end, Lee conceded that the only policy that PMSI* might justify is nonextended counter-city nuclear deterrence, but he stressed this justification depends (as it did for Dummett) on a certain knowledge that nuclear threats alone would cause a deterree to abandon their plans of aggression.[121] Lee's account suggests that even nonextended (or national) counter-city nuclear deterrence is not justified, since each nuclear-armed great power could reasonably find a substitute in conventional deterrence. It might be objected that, in comparison to the relatively cheaper costs of nuclear deterrence, the alternative of an enhanced conventional deterrent would come at the unacceptable expense of the kind of economic progress that the United States and many U.S. allies experienced throughout much of the Cold War era. To my knowledge, these are not objections that Lee, or the other Kantian-inspired commentators, chose to address. However, their accounts suggest that the greater expense of a conventional deterrent would have been morally preferable if it could secure the survival of humanity (and one's own state) while not putting innocents at risk of undue harm.

THE MORAL IMPERATIVE OF THE GREATER GLOBAL GOOD: ANTI NUCLEAR CONSEQUENTIALISM

The last ethical approach to review on the questions of nuclear deterrence and reprisal policy is a clear contrast to the rationalist consequentialism of Gauthier and Kavka and the hybrid consequentialism of Nye. This anti-nuclear consequentialism rejected as its foundational principle the zero-sum pursuit of national security in favor of the utilitarian principle of the greater good for the greatest number of relevant or affected persons. Like the rationalists, the utilitarian commentators were concerned about questions of risk and probability in nuclear defense and deterrence policy. However, their cosmopolitan commitments led to sharp disagreements with the rationalists on the nature of the risks of nuclear deterrence failure leading to nuclear warfare. In the end, the utilitarians were confident in their pessimism of nuclear deterrence's efficacy.

For instance, Todd Gitlin urged the superpowers to "move beyond deterrence."[122] Gitlin acknowledged that "deterrence has, so far, not failed"; however, his concern about deterrence failure was provoked by the early Reagan administration's reemphasis on a nuclear warfighting strategy and an augmented nuclear capability on the grounds that "the ability to 'prevail' in a nuclear war is necessary for deterrence." As Gitlin stated,

> The flaw in the doctrine of deterrence is that, to work, it has to work perfectly, and it has to work forever. Yet, we are talking about a human process; people miscalculate. Nations blunder into war. In their terror and myopia, the superpowers have adopted a system that can succeed only if it works indefinitely longer than any social arrangement in history ever has.[123]

Thus, Gitlin framed the question of deterrence in all-or-nothing terms. This is to say, while some deterrence failures might not lead to catastrophic nuclear outcomes, we could never know if the first failure, or any subsequent failures, would actually avoid catastrophe. For this reason, we become compelled to hope against reason that nuclear deterrence will work "indefinitely longer" than has any other security policy. Unfortunately, the institution of nuclear deterrence is vulnerable to human imperfection in all its variations. Thus, Gitlin argues that

> faith in perpetual deterrence . . . requires forgetting what we know about stupidity, error and evil in high and middling places. Although the timing of deterrence's failure cannot be predicted with certainty, sooner or later, by design or miscalculation or sheer insanity at the top, some of the bombs could go off. When they do, it won't be much consolation knowing that for a brief span of human history, deterrence worked.[124]

In short, Gitlin could not foresee any likely future condition where nuclear deterrence did not succumb to the malice or fallibilities of the leaders or officers of nuclear-armed states. The sanguine attitudes of deterrence advocates rested on a critical misunderstanding of the stakes involved. For Gitlin, committing to nuclear deterrence is like playing Russian roulette, even if the revolver had 100 chambers.[125] The only morally responsible approach to nuclear policy would thus be nuclear abolition.

Nye did not find Gitlin's argument compelling.[126] For one, Nye claimed that Gitlin did not adequately distinguish between nuclear deterrence failures, which would lead to nuclear war, and those which might lead to the strengthening of nuclear nonuse norms by augmented international mechanisms. Thus, Gitlin's metaphor of playing Russian roulette with a 100-chamber revolver was misconceived, since the odds of nuclear war involved a series of "independent and interdependent probabilities" each of which drove the odds of nuclear war lower than higher.[127]

A contrasting anti nuclear consequentialist view with Gitlin's came from Robert Goodin, who contended that the probabilistic approach to questions of nuclear deterrence success or failure did not make sense because

> we just do not know enough about the shape of the underlying distribution to justify employing any of the standard techniques for estimating probabilities. . . . The balance of terror has kept the peace for the past thirty-five years, to be sure. But thirty-five years is just too short a run on which to base our probability judgments, given the unacceptability of even very small probabilities of such a very great horror [as all-out nuclear war].[128]

For Goodin, the brief period of Cold War history (which some historians and strategists described as a "long-peace"[129]) did not offer the kinds of variation in nuclear- deterrence-related behavior necessary for probability estimation. Hence, he claimed that "it is altogether inappropriate to engage in probabilistic reasoning about the chances of a breakdown in the balance of terror that leads to large-scale nuclear war."[130] It is inappropriate and morally irresponsible precisely because it amounts to "playing the odds [with nuclear deterrence] without knowing the odds. That constitutes recklessness par excellence."[131] The morally responsible alternative, for Goodin, is to adopt "possibilistic" reasoning limited to three categories: the impossible, the possible, and the certain.[132] For any policy question where we lack adequate knowledge of the underlying distribution, such as the questions of nuclear deterrence and war, it is thus morally imperative to make good outcomes certain and make bad outcomes impossible.[133] In turn, this requires action to make nuclear disarmament certain. In a world of two nuclear-armed superpowers, bilateral nuclear disarmament is preferable; however, unilateral nuclear disarmament is morally necessary to prevent an all-out nuclear war.

Disarmament skeptics might raise two objections to Goodin's argument: (1) the disarmed power is unacceptably vulnerable to conquest by the rival nuclear superpower, and (2) this rival superpower might be ruled by a mad and genocidal tyrant.[134] To the first objection, Goodin argued that the chances of nuclear aggression were virtually impossible if the sole remaining nuclear superpower was minimally rational and pursued the customary goals of world politics. For instance, no minimally rational aggressor would destroy the spoils of war they sought to enjoy, and the use of nuclear weapons by one superpower against a nuclear-free one could not avoid destroying the spoils of war. Consequently, the newly emergent nuclear-free superpower could achieve effective deterrence by fielding an adequately robust, conventional military force.[135] In this regard, Goodin's position recalls Lee's advocacy of conventional deterrence. Unlike Lee, however, the reason for choosing the conventional alternative is not about avoiding nuclear hostage-holding; it is about making nuclear war impossible by eliminating one of its necessary components. To the second objection, Goodin argued that the Western experience with mad and genocidal tyrants was such that we could identify and deal with them "long before they have a chance to make any real trouble for us."[136] And even if a recently denuclearized superpower was too slow on this point, we could still rebuild the nuclear deterrent and return to the *status quo ante* balance of terror.

When assessing Goodin's two responses, this latter one seems the least convincing. Indeed, during a conference in which Goodin's argument was presented, the strategist Colin Gray observed that "we have not been very good at predicting crazy leaders very far in advance" and "it may take a decade or more to reorient our defensive posture completely" to address a new and tyrannical nuclear threat.[137] Goodin did not respond to the Gray's first rejoinder, but to the second one Goodin doubted that deterrence of any kind would affect the calculations of a truly mad and genocidal tyrant. Goodin didn't seem to notice, however, that Gray's second rejoinder related a sufficient consequentialist reason for each superpower to retain their nuclear arsenal in the expectation that a mad and genocidal tyrant might someday take control of a superpower. If Gray is correct on this point, then Goodin's larger anti nuclear argument is significantly weakened.

CONCLUSION: THE MORAL DILEMMAS
OF NUCLEAR POLICY

This chapter has argued that the lack of consensus among nuclear ethicists on the morality of nuclear defense and deterrence policy was based on a series of fundamental disagreements over which moral principles had privileged or non-overridden status. These principles were, respectively, the *jus ad bellum* rights of states or peoples to survival and security (or the preservation of their way of life) *as opposed to* an absolute *jus in bello* right of noncombatants to survival and security *and/or* the consequentialist moral duty to realize the global greater good and the survival of humanity. In the absence of any agreement on the ranked order of these values, the nuclear ethical debates were all but guaranteed to not find consensus.

The chapter also related commentators' agreement on the immorality of "all-out" nuclear war, but its significance is not entirely clear. On the one hand, this agreement seems entirely banal (well, of course, the annihilation of humanity is evil!); on the other hand, it seems supremely important as a background condition for the justification or condemnation of nuclear defense and deterrence policies that rely on raising the risks of all-out nuclear war in the name of national security. Additionally, commentators agreed generally on the evil of risking noncombatant deaths by nuclear deterrence policy. They disagreed, however, on whether the principle should be understood as absolute and non-overriding. In the end, their principled disagreements rendered the few points of agreement they had as largely nonconsequential.

The upshot is that the Cold War nuclear ethics debate did not resolve the core moral dilemmas concerning nuclear defense and deterrence policy.[138] One unresolved dilemma might be called the "contradictory moral rule dilemma"[139] that arises when the moral imperative to prevent nuclear catastrophe generates contradictory maxims for state security policy: (1) commit to counter-city nuclear deterrence and nuclear reprisal if necessary *and* commit to counter-city nuclear deterrence but avoid nuclear reprisal or (2) commit to limiting nuclear deterrence and reprisal to counterforce targets *and* commit to making nuclear disarmament into a moral certainty. However, the chief moral dilemma left unresolved is that of "competing moral requirements." In the absence of a (psycho)logically compelling principle that might reconcile the moral imperatives of state security, human rights, and the greater global good, the field of Cold War–era commentators was not able to construct a principled consensus view. This was the case, despite Nye's attempt to forge a realist-cosmopolitan synthesis in which the moral considerations of motive, means, and consequences were each given their proper due.[140]

The remainder of this book turns to critically examine the morality of twenty-first-century nuclear defense and deterrence policy as it has evolved from its Cold War origins. Chapter 3 examines contemporary nuclear defense and deterrence policy in light of a moral imperative of common security, an idea introduced during the Cold War but which has been radically underappreciated. Chapter 4 examines the moral incoherence of liberal democratic states, whose commitments to nuclear deterrence constitute contradictions for their ontological security. Chapter 5 examines the question of nuclear disarmament in light of recent expressions of the Humanitarian Imperative and the contrasting position of "Conditions-Focused" nuclear disarmament. Several themes in this chapter and in chapter 3 will recur in chapter 5. Chapter 6 concludes the book by drawing inferences from the aforementioned chapters concerning a twenty-first-century nuclear ethics resting on the priority of just orders in the production of a secure world of states and peoples.

NOTES

1. National Conference of U.S. Catholic Bishops (1983, para. 122).
2. Walzer (2015).
3. Kant (1795, 1996d).
4. Scholars disagree on the proper periodization following the atomic attacks against Hiroshima and Nagasaki. Some prefer to classify the entire historical period from 1945 to the present as a singular nuclear age while others believe dividing the period into first and second nuclear ages (or even more) is more appropriate. For the former view, see, for example, Gavin (2012). For the latter, see, for example, Solingen (2007). This chapter favors the former view.
5. National Conference of U.S. Catholic Bishops (1983, para. 122).
6. Walzer (2015).
7. Bok (1988).
8. J. Robert Oppenheimer was the director of the Los Alamos Laboratory that, under the aegis of the Manhattan Project in the United States, developed the first atomic bomb.
9. Ham (2011, 464). For a broader account of the moral numbness introduced by the use of "terror bombing" in World War II, see Ellsberg (2017).
10. For a selection of accounts that examine the history of the Cold War with a focus on nuclear statecraft issues, see Gavin (2012); Smoke (1993); Walker (2012).
11. Ambrose and Brinkley (1997, 181–86).
12. Hoffman (2009, 94–99); Walker (2012, 101).
13. For an account of Christian apocalyptic perspectives on nuclear warfare, see Cook (2004).
14. For competing accounts of why nuclear weapons have not been used since 1945, see Paul (2009) and Tannenwald (2007).
15. The limitation of this chapter to English-speaking commentators is not intended to portray debates found in non-English literatures as unimportant. To the contrary, understanding these literatures is essential to gain a cosmopolitan understanding of nuclear ethics as it has evolved over time. Rather, the limitation of this chapter speaks more of my own language limitations than anything else.
16. Walker (2012).
17. For other post–Cold War accounts of the morality of nuclear deterrence and reprisal policy, much of which focuses on Cold War sources, see, for example, Hashmi and Lee (2004); Heinze (2016); Lee (1993); Orend (2013, chapters 4–7).

18. See, for example, Elshtain (1985); Tickner (1991). More recently, see, for example, Buzan and Hansen (2009, 138–41); and Sjoberg (2010).

19. See, for example, Nakamitsu (2018) and Intondi (2015). See also Wittner (2009).

20. For extensive discussions of deterrence theory, see, for example, Morgan (2003); Schelling (1960, 1966); and Quinlan (2009). For an alternative theoretical perspective, see Vuori (2016).

21. See, for example, Kavka (1978); Gauthier (1984).

22. Walker (2012, 60).

23. See Morgan, (2003, chapter 4).

24. See especially Walker (2012, 24, 84, 123).

25. The term "non-overridden" refers to a moral principle or value that cannot be superseded by a competing principle or value. The term is borrowed from Sinnott-Armstrong (1988).

26. Nye (1986, 43).

27. Anscombe (1961); National Conference of U.S. Catholic Bishops (1983); Hehir (1986); Quinlan (1981 (2009)); Payne and Payne (1987). For a thorough examination of Quinlan's arguments, see also Jones (2013) and Ogilvie-White (2011).

28. Boyle (1992); National Conference of U.S. Catholic Bishops (1983, para. 72); Payne and Payne (1987, 46).

29. National Conference of U.S. Catholic Bishops (1983, para. 96).

30. National Conference of U.S. Catholic Bishops (1983, paras. 60, 85).

31. "St. Paul's Epistle to the Romans" (2011, chapter 13); Anscombe (1961, paras. 30–54); Payne and Payne (1987, 46).

32. Walzer (2015 (1977), 53–54).

33. Walzer (2015 (1977), 58–62).

34. Gauthier (1984, 494–95).

35. See also Kavka (1978, 287).

36. National Conference of U.S. Catholic Bishops (1983, para. 101); Orend (2013, 119–21).

37. National Conference of U.S. Catholic Bishops (1983, para. 99); Orend (2013, 125–26).

38. See, for example, Kierulf (2017, 13–17, 160).

39. Smoke (1993, chapters 3 and 4).

40. Jones (2013); Quinlan (2009, 181).

41. Payne and Payne (1987, 32).

42. Gauthier (1984, 482).

43. Payne and Payne (1987, 53, 101–6); Quinlan (2009 (1984), 186); Nye (1986, 45).

44. Mitzen (2006b); Steele (2008, 1–7). For a distinction between egalitarian-liberal and communitarian traditions within Western thought, see the relevant sections in Farrelly (2004).

45. Nye (1986, 45).

46. Nye (1986, 45).

47. Nye (1986, 61–70).

48. Nye (1986, 66–67).

49. I borrow the term "entrenchment" from William Walker, who uses this concept to explain in part the resilience of the international order of nuclear restraint. See Walker (2012, 13–14).

50. Nye (1986, 78–79).

51. Nye (1986, 115–20).

52. Gauthier (1984, 482).

53. Gauthier (1984, 479–80).

54. Gauthier (1984, 489).

55. Kavka (1978, 288).

56. Kavka (1978, 291). Emphasis in the original.

57. I thank Harry Gould for these questions.

58. See, for example, Austin (1975); Buzan and Hansen (2009, 212–15); Buzan, Waever, and Wilde (1998); Fierke (2015, 110–30).

59. Kavka (1978, 292).

60. Kavka (1978, 294).

61. Kavka (1978, 295).

62. See Orend (2013, 164–66).

63. Anscombe (1961); Ramsey (1962, 145–47).

64. Ramsey (2002 (1968), 171). For critiques of Ramsey's baby-on-front-bumpers analogy, see, for example, Nye (1986, 93); Walzer (2015 (1977), 269–70).

65. Ramsey (1962, 161–70).

66. Payne and Payne (1987, 125–34).

67. See Orend (2013, 166–67).

68. Hehir (1975, 58).

69. Mearsheimer (2014, 224–30); Ellsberg (2017, Introduction).

70. Walzer (2015 (1977), 272).

71. Hehir (1975, 69).

72. National Conference of U.S. Catholic Bishops (1983, paras. 147–48, 73).

73. Quinlan (2009, 51).

74. Nye (1986, 53–54).

75. Nye (1986, 54).

76. Orend (2013, 121). See also Nye (1986, 55–56). For a recent critical view of the mainstream understanding of double effect, see Gould (2013).

77. Quinlan (2009, 52).

78. Quinlan (2009, 185–86).

79. Anscombe (1961, 51–52).

80. Anscombe (1961, 51–52).

81. Walzer (2015 (1977), 272–74).

82. Walzer (2015 (1977), 250–54).

83. Walzer (2015 (1977), 274).

84. Walzer (2015 (1977), 282).

85. Walzer (2015 (1977), 282–83). Also, Quinlan (2009, 54–55).

86. The Gospel of Matthew 6:33, *The Bible*, King James Version, accessed on August 16, 2018, at https://biblehub.com/kjv/matthew/6.htm.

87. Bok (1988, 9).

88. Kant (1996b, 73–80).
89. My treatment of these two formulations is influenced by Korsgaard (1996, chapters 3 and 4).
90. Kant (1996b, 73).
91. Kant (1996b, 80).
92. Kant (1784 (1996a)).
93. For a direct statement on this, see Shue (2004, 156–57).
94. Donaldson (1985, 540).
95. Donaldson (1985, 541).
96. Quinlan (2009, 48).
97. Quinlan (2009, 186).
98. Nye (1986, 51–52).
99. Donaldson (1985, 545–46).
100. Donaldson (1985, 547).
101. Dummett (1986, 112).
102. Dummett (1986, 112).
103. Dummett (1984, 33). Also quoted in Nye (1986, 11).
104. Nye (1986, 11).
105. For an extensive discussion of international humanitarian law and its relation to nuclear weapons policy, see Kierulf (2017).
106. Nye (1986, 11).
107. On the epistemological question of conceptual entailments, see Brandom (2000).
108. Dummett (1986, 122).
109. Booth and Wheeler (2008, 21–41).
110. Dummett (1986, 125).
111. Lee (1985, 549). On this point, Lee finds a partial consensus with the approaches of the just war commentators, whose *jus ad bellum* and *jus in bello* principles incorporated consequentialist and nonconsequentialist principles, as well as Nye who had explicitly called for a three-dimensional ethical approach to the questions of nuclear policy. See, for example, National Conference of U.S. Catholic Bishops (1983, paras. 80–99); Nye (1986, 20–26).
112. Lee (1985, 550).
113. Lee (1985, 551).
114. Lee (1985, 552–53).
115. Lee (1985, 553–56).
116. Lee (1985, 556).
117. Lee (1985, 558).
118. For a quantitative analysis of U.S.-Soviet extended nuclear deterrence policies, see Weede (1983).
119. Lee (1985, 561).
120. Lee (1985, 564).
121. Lee (1985, 565–66)
122. Gitlin (1984).
123. Gitlin (1984).

124. Gitlin (1984).

125. Gitlin (1984).

126. Nye (1986, 62–65). For other antinuclear consequentialist accounts on the near-inevitability of deterrence failure and nuclear war, see, for example, Lyttle (1983); Schell (1982); Lackey (1984).

127. Nye (1986, 66–67). See earlier discussion on the points of general consensus among pro-deterrence commentators on the question of deterrence as a risk-reward strategy.

128. Goodin (1985, 643).

129. Gaddis (1989).

130. Goodin (1985, 644).

131. Goodin (1985, 644).

132. Goodin (1985, 645).

133. Goodin (1985, 649).

134. Quinlan (2009, 50–51).

135. Goodin (1985, 655).

136. Goodin (1985, 656).

137. Goodin (1985, 656, n. 37).

138. For a more detailed account of the moral dilemmas of nuclear weapons policy, both during and after the Cold War, see Doyle II (2015b).

139. I owe the names of the moral dilemmas cited to Sinnott-Armstrong (1988).

140. Nye (1986, chapters 2 and 7).

Chapter 3

Nuclear Defense and Deterrence in the Twenty-First Century and the Moral Imperative of Common Security

The previous chapter explored the Cold War–era debate among nuclear ethicists on the morality of superpower nuclear defense and deterrence policies. It contrasted arguments defending these policies based on the moral imperatives of national survival and self-defense[1] with critical arguments anchored on the moral imperatives of human rights and the greater global good.[2] It noted a key intramural dispute among nuclear defense and deterrence advocates on whether *jus in bello* imperatives of noncombatant immunity and proportionality imposed an absolute side constraint on policies of limited nuclear warfare. On the one hand, the U.S. Catholic Bishops argued that *jus in bello* imperatives forbade the use of nuclear weapons, which led them to justify nuclear deterrence but not nuclear use. The contrasting view of Sir Michael Quinlan, a former British defense official who also was Catholic, held that *jus ad bellum* requirements of self-defense and survival were absolute (at least, for Western democracies) and therefore the defensive use of nuclear weapons must be morally permissible if deterrence was morally justifiable. In the end, the moral debate on nuclear defense and deterrence did not overcome such deadlocks by the end of the Cold War era, leaving the post–Cold War world without an undisputed set of theoretical or practical resources from which fresh thinking might be generated on the global nuclear dilemma.

Theoretically, this deadlock or dissensus exposed the absence in nuclear ethical analysis of adequate bridge principles that might reconcile competing moral imperatives *or* offer a noncontroversial ranking of moral requirements to resolve the moral dilemmas of nuclear defense and deterrence.[3] One political implication of this dissensus was that it enabled a confident moral justification of nuclear defense and deterrence policies by state officials, policy experts, and scholars in the nuclear-weapon states (NWS). In turn, it meant that their nuclear defense and deterrence policies were implemented without effective moral opposition and that nuclear abolitionist views were increasingly marginalized,[4] even though antinuclear movements occasionally pressured U.S. and Soviet officials to undertake modest nuclear freezes or cuts.[5] Consequently, the moral debate on nuclear defense and deterrence remains in its Cold War impasse to this day, even though a series of significant structural changes in the international order occurred during the post–Cold War era. These changes will be examined more closely in the next section.

As the first chapter argued, the endurance of this nuclear *status quo* suggests the need for a new nuclear ethics more appropriate for early twenty-first-century political conditions. To think nuclear ethics afresh, a suitable alternative anchoring moral principle must be theorized on which a new set of nuclear ethical maxims might be fixed. Accordingly, this chapter explores the possibilities and limits of such an anchoring principle by way of two interrelated Cold War–era accounts. The first is Karl Deutsch's conception of "security communities," which was derived from several modern case studies of security cooperation among North Atlantic countries.[6] The second is Olof Palme's conception of "common security," which, as the former Swedish prime minister, he introduced to address the moral and political dilemmas of small states that were vulnerable to the effects of the mutual hostility of the superpowers.[7] Each of these conceptions contains a set of rudimentary moral intuitions from which new thinking on nuclear ethics might develop.[8]

This chapter's main argument is that the Deutsch and Palme accounts identify a moral imperative of common security that can resolve or harmonize the competing imperatives of national and human survival and security at the heart of our persistent nuclear dilemma. This common security imperative can constitute a bridge principle for nuclear ethics lacking in the Cold War–era debates and which can anchor a twenty-first-century nuclear ethics. This chapter is limited to explicating this common security imperative, the applications of which are explored in the book's remaining chapters.

Accordingly, this chapter's first section briefly describes the evolution of nuclear defense and deterrence policies since the Cold War's end, focusing mostly on the U.S.-Russian relationship as Cold War bipolarity was replaced by post–Cold War unipolarity. And while this brief description cannot replace a more thorough account of post–Cold War nuclear politics, it is tailored to highlight the relevant contrasts between the Cold War and post–Cold War eras for later nuclear ethical analysis. In the chapter's second section, the Deutsch and Palme accounts on security communities and common security are reviewed, highlighting their innovative contributions and their theoretical and practical limits. These two lengthy sections are necessary stage-setting pieces for advancing the explication and defense of the common security imperative in the last two sections.

NUCLEAR DEFENSE AND DETERRENCE IN THE TWENTY-FIRST CENTURY

The evolution of U.S. nuclear defense and deterrence policies—and those of other NWS reacting directly to U.S. nuclear policy—during the post–Cold War era (*c.* 1992–present) appeared to correspond to significant large-scale shifts in the international order. The first was from the Cold War bipolar configuration (i.e., the United States and the former USSR as the rival "poles" of power) to the unipolar configuration of the early post–Cold War era, where the United States emerged as the world's only superpower. The second shift was from this "unipolar moment"[9] into that which exhibited greater degrees of multipolarity or polycentricity.[10] While the beginning of this unipolar moment is relatively undisputed (i.e., the early 1990s), it is difficult to determine exactly when it succumbed to an emergent but not fully realized multipolarity. The U.S. Defense Department's 2018 Nuclear Posture Review identifies 2010 as the year of a clear resurgence of great power politics, implying a shift to multipolarity.[11] A contrasting analysis related later in the chapter suggests that this shift began in the early to mid-2000s and became increasingly evident as (1) Russia successfully attacked and annexed a portion of Georgian territory in 2008 and Ukraine in 2014, (2) China built and then militarized islands in the South China Sea, and (3) North Korea initiated a nuclear testing program between 2006 and 2017, all the while remaining resilient (and even experiencing some economic growth) under U.S.-led United Nations sanctions. The next subsection describes in more detail the evolution of U.S. nuclear defense and deterrence policies that corresponded to these shifts.

The Evolution of Nuclear Deterrence: From Bipolarity to Unipolarity

The Cold War era was characterized largely by a bipolar distribution of power between two rival superpowers, the United States and the former Soviet Union.[12] In the early 1990s, the Soviet Union disintegrated into fifteen new and independent states, with the newly formed Russian Federation inheriting the former Soviet nuclear arsenal and its permanent place on the United Nations Security Council.[13] The fundamental diminishment of Russia's national status within the world order meant that Cold War bipolarity had been replaced by post–Cold War unipolarity with the United States as the sole superpower. Moscow could no longer project conventional military or economic power beyond its borders, even though it retained a larger nuclear arsenal than did Washington.[14] Moreover, Moscow's partial transition from authoritarian to democratic rule in the 1990s offered hope to many Westerners that its ideological enmity with the United States had come to an end.

Table 3.1   A Reconstruction of Competing Cold War Nuclear Ethical Approaches*

Nye's Five Maxims of Nuclear Ethics	Quinlan's Just War Stance	U.S. Catholic Bishops' Just War Stance	Walzer's Just War Stance	Kantian Nuclear Ethics (Donaldson, Dummett, Lee)
Self (national)-defense is a just but limited right.	National defense for liberal states (which especially protect religious liberty) is an absolute and non-overridden right.	National defense is a just but limited right in relation to the value of God's creation and *jus in bello* constraints.	National defense is a near-absolute and near-non-overridden right of such a community (i.e., supreme emergency condition).	Self (national)-defense is a just but limited right.
Never treat nuclear weapons as normal weapons.	Acknowledge that nuclear weapons are not "normal" but they are necessary.	Never treat nuclear weapons as normal weapons.	Nuclear weapons are not normal weapons.	Do not rely on technologically recalcitrant weapons systems like nuclear weapons.
Minimize harm to innocent people.	Risks to innocent people in rival states are regrettable, but not worse than capitulation to hostile and illiberal state-enemy.	Nuclear harm (and the threat of it) is immoral on *jus in bello* grounds.	Risks to innocent people are regrettable, but it seems justified in cases of supreme emergency.	Do not commit mass murder, and do not threaten to commit mass murder, to achieve a goal or establish a condition politically preferable to oneself (or country).
Reduce the risks of nuclear war in the short term.	Reduce the risks of nuclear war by maintaining a robust and credible nuclear deterrent.	Reduce the risks of nuclear war in the short term.	Reduce the risks of nuclear war by maintaining a credible nuclear deterrent.	Do not systematically violate the fundamental rights of any innocent person(s).
Reduce reliance on nuclear weapons over time.	Reduce reliance on nuclear weapons only if existential threats against liberal states disappear.	Reduce reliance on nuclear weapons over time, and work in good faith toward nuclear disarmament.	Reduce reliance on nuclear weapons only if existential threats against one's (liberal) state disappear.	Do not rely on policies of threats of vicarious punishment.

* Table 3.1 is a summary of the review of Cold War era approaches to nuclear ethics. See chapter 2 for that discussion.

This shift to unipolarity had and still has significant effects on the evolution of U.S. nuclear defense and deterrence policies. Table 3.1 summarizes this evolution through the beginning of the Trump administration.[15]

Table 3.1 recalls chapter 2's discussion that Cold War–era nuclear defense and deterrence postures were ultimately defined by the concept of "mutually assured destruction" (MAD). On this concept, the superpowers were mutually committed to nuclear reprisal in the case of major deterrence failure (e.g., a Soviet invasion of Western Europe). Although strategists emphasized that U.S. and Soviet leaders in such cases might limit the number or intensity of nuclear reprisal strikes, many others feared that unrestrained nuclear war would inevitably follow and lead to the virtual or actual destruction of the superpowers and other countries within reach of nuclear blast, fire, and fallout.

Table 3.2 thereafter relates that, following the Soviet Union's disintegration, the George H. W. Bush administration (Bush 41) embraced the hope of an enduring peace with Russia and, after an initial pause and moment of recalculation, introduced modifications to its long-standing security policies.[16] One key change was the demotion of nuclear deterrence policy in favor of cooperative threat reduction, which involved a series of unilateral actions (which followed Gorbachev's own policies of nuclear demotion) and

Table 3.2   The Evolution of U.S. Nuclear Deterrence and Defense Policy through Obama

U.S. Administration	U.S. Nuclear Deterrence and Defense Policy
Cold War era	Mutually assured destruction (MAD); prevailing strategy in case of deterrence failure
Bush 41, 1989–1993	Cooperative threat reduction: unilateral nuclear arms reduction; de-alerting and de-targeting—a shift to general deterrence from existential deterrence
Clinton, 1993–2001	Mutually assured safety: continuation of threat reduction but hedge against resurgent Russia; tailored deterrence to address threat of "rogue" state uses of biological and chemical weapons
Bush 43, 2001–2009	Prevention of terrorism, nonstate and "rogue" state uses of weapons of mass destruction (WMD), and rogue-state proliferation; augment ballistic missile defense against "rogue" state threats
Obama, 2009–2017	Strategic stability; nonnuclear strategic deterrence; extended nuclear deterrence; tailored deterrence; ballistic missile defense against "rogue" state threats
Trump, 2017–	Deter adversary nuclear and nonnuclear attacks against the United States and its allies; tailored deterrence and flexible approaches (i.e., low-yield options) across a spectrum of threats and contexts; nuclear defense in extreme circumstances, including adversary nuclear and nonnuclear attacks and nuclear terrorist attacks

a series of bilateral and multilateral actions. Unilaterally, Bush 41 ordered the de-alerting of U.S. strategic bombers, the de-targeting of Russian and Warsaw Pact military assets and cities, the de-alerting of nuclear-tipped intercontinental ballistic missiles scheduled for deactivation, the termination of other nuclear modernization programs that were before Congress, and the withdrawal of nonstrategic nuclear weapons (NSNW) from U.S. Navy surface ships, submarines, and land-based aircraft. These orders replaced MAD with a recessed generalized deterrence policy in which the United States did not target any state or asset specifically.[17] Bilaterally and multilaterally, Bush 41 worked to reduce the risks of nuclear terrorism and "rogue" state nuclear proliferation by encouraging Moscow to accept Washington's help to increase the physical security of their nuclear weapons and energy facilities. It also insisted on strengthening the Nuclear Nonproliferation Treaty (NPT) regime's regulatory provisions to prevent an increased threat of illicit nuclear transfers, and it intensified the U.S. commitment to international institutions in the realms of security and trade that, it was believed, would reinforce the structures of international stability and security.

As the transition into the post–Cold War era continued, the Clinton administration followed and expanded Bush 41's institutionalist approach under the rubric of "mutual assured safety." The aim was to continue with nuclear reductions but retain enough nuclear deterrent capability in case Russia returned to a Cold War–like hostility against the United States or in case new nuclear proliferators succeeded in acquiring nuclear weapons to further their revisionist goals. Clinton also introduced a policy later known as "tailored deterrence," which contrasted with previous administrations' "one size fits all" approaches. In the new unipolar world, Clinton administration officials conducted a vigorous debate on the role of nuclear weapons and eventually came to the position that nuclear defense and deterrence policies needed to be shaped to fit the unique geopolitical conditions of each region of concern. In the wake of Iraqi leader Saddam Hussein's use of chemical weapons in 1988 against his Kurdish population, one enduring application of this tailored approach was to use nuclear threats to deter future "rogue" state chemical weapons attacks.[18] Despite some concerns over this use of nuclear deterrence, it became a standard feature of subsequent administrations' security policy.[19]

The next significant moment in the evolution of U.S. nuclear defense and deterrence policy occurred in the wake of the profoundly shocking September 11, 2001, terrorist attacks against New York and Washington, D.C. The George W. Bush administration (Bush 43) perceived these attacks as confirmation that Islamic terrorist organizations and their state sponsors were not deterrable and, accordingly, elevated the prevention of nuclear terrorism higher on the U.S. national security agenda. Moreover, Bush 43 was suspicious of institutionalist approaches, preferring a relatively unrestrained unilateralism over the slower and more complicated process of working with allies and the UNSC. Consequently, its 2002 National Security Strategy (NSS 2002) reapplied the Clinton administration's commitment to address the threats "at the crossroads of radicalism and technology."[20] Bush 43's nuclear deterrence policy remained largely general and recessed, partly as an assurance policy for allies and partly as a reserve capacity to hedge against a future resurgent Russia. However, Bush 43's withdrawal from the 1972 Anti-Ballistic Missile Treaty and his subsequent commitment to augmenting missile defense systems helped, along with NATO expansion in Eastern Europe (which began in the Clinton era), to activate Russian fears about the continued extension of U.S. power toward Moscow's near-abroad. Unfortunately, U.S. failure to effectively address these fears became a factor in Russia's 2008 incursion into Georgia, their 2014 annexation of Crimea in Ukraine, and the subsequent increase in U.S.-Russia tensions, each of which became early indicators of the shift from unipolarity to multipolarity or polycentricity in the international order.[21]

After the 2008 U.S. presidential election, the Barack Obama administration rolled back Bush 43's emphasis on preventive war and regime change and, as a partial result, "reset" U.S. relations with Russia back toward increased security cooperation. Obama believed that a successful reset would maintain the recessed and general nature of U.S. nuclear deterrence policy and help to lay the groundwork for an eventual nuclear abolition. Indeed, Obama's 2009 Prague Speech signaled that the new administration believed nuclear disarmament could be put on hold indefinitely, and among global civil society groups and abolitionist NNWS hopes were raised significantly that the United States (and, in turn, the other NPT NWS or N5 states) would honor their NPT commitments to take concrete actions toward nuclear disarmament, such as ratifying the Comprehensive Nuclear-Test-Ban Treaty.[22]

Accordingly, Obama's Nuclear Posture Review (NPR 2010) specified several goals: strengthen nuclear nonproliferation efforts, strengthen efforts against nuclear terrorism, reduce the role of nuclear weapons in U.S. strategic planning, ensure strategic stability during the ongoing process of nuclear reductions, strengthen extended deterrence and assurance, and maintain a safe and secure deterrent as long as other states kept their nuclear weapons.[23] Moreover, the Obama administration sought to align its nuclear defense policy more strictly with the Law of Armed Combat and the *jus in bello* principles of discrimination and proportionality.[24] These policy shifts, combined with strong rhetorical commitment at the 2010 NPT Review Conference to previous U.S. disarmament commitments, offered hope to many NPT member states and civil society groups that the evolution of nuclear defense and deterrence policy would eventually culminate in a global and irreversible nuclear disarmament.[25]

The Evolution of Nuclear Deterrence: From Unipolarity to Multipolarity/Polycentricity

By 2014–2015, Washington faced two security dilemmas of increasing intensity. One involved the attempts of Russia and China to reassert themselves as great powers. Each sought greater influence, if not hegemony itself, in their respective regions to escape perceived humiliation suffered at the hands of U.S. expansionistic policy from 1990 onward and to (re)claim their status as great powers of the highest rank. In Europe, Russia annexed Crimea in 2014, intensified submarine patrols in the Baltic Sea, modernized its nuclear forces, and began to capitalize on U.S. and European domestic divisions by means of misinformation campaigns in Western social media and election interference in NATO countries. For its part, China gradually expanded into and militarized the South China Sea, began its own nuclear modernization program, and aggressively began a campaign of cyberattacks against U.S. political and economic assets.[26]

Other new security challenges for the United States were the increased prospects of a nuclear-armed North Korea and Iran to effectively use weak-state deterrence to balance against U.S.-backed regional security orders in East Asia or the Middle East.[27] By 2013, the North Korean government had completed three nuclear test detonations and began an aggressive ballistic missile testing program, despite suffering the imposition of economic sanctions by the UNSC.[28] Fearing the same prospect in the Middle East, the Obama administration, along with Russia, China, Britain, France, and Germany (the P5 + 1), had begun negotiations with Iran on an agreement to curtail the latter's nuclear program in exchange for unfreezing Iranian assets in the United States and other kinds of economic assistance. By 2015, these efforts succeeded despite fierce opposition from Israel and hardline conservative voices in Iran and the United States.[29] The final agreement was called the Joint Comprehensive Plan of Action (JCPOA), and it remained unchanged for two-and-one-half years until the Trump administration unilaterally withdrew from it in early 2018.[30]

Despite these new security challenges that permitted U.S. policymakers to support the continuation of nuclear defense and deterrence, abolitionist NNWS were increasingly disenchanted with the progress the United States and other NWS had made toward nuclear disarmament. They were dismayed at the published reports of U.S. nuclear modernization, and they increased pressure on NPT member states to produce a treaty banning nuclear weapons.[31] Their efforts during the 2015 NPT Review Conference were resisted firmly by the NWS and their allies (see chapter 5). Believing that the antinuclear community and the U.S. public generally did not appreciate the dangers of nuclear disarmament, Brad Roberts wrote a book-length defense in 2016 of the necessity for U.S. nuclear defense and deterrence going forward.[32]

Roberts, a former Deputy Assistant Secretary of Defense for Nuclear and Missile Defense Policy in the Obama administration (2009–2013),[33] contended that nuclear disarmament was dangerously risky if regional challengers (e.g., Russia, North Korea) retained or were actively seeking nuclear weapons for themselves. Roberts recalled the 2009 Commission on the Strategic Posture of the United States report, which said in part that "the conditions which might make possible the global elimination of nuclear weapons are not present today and their creation would require a fundamental transformation of the world order."[34] Instead of accepting any false hope that nuclear disarmament by itself would facilitate peaceful relations among historic state rivals, Roberts argued that the United States must first understand how challengers would use nuclear capabilities to secure political victories and afterward devise strategies to effectively counter their efforts.

In making this argument, Roberts related a core insight in Paul Nitze's metaphor of the game of chess with an "atomic queen." Nitze had anticipated that

> the atomic queens [of regional challengers] may never be brought into play; they may never actually take one of the opponent's pieces. But, the position of the atomic queens may still have a decisive bearing on which side can safely advance a limited-war bishop or even a cold war pawn.[35]

With Nitze's metaphor of the atomic queen in mind, Roberts distinguished between three types of nuclear defense and deterrence strategies that North Korea might assert to prevent a U.S. invasion or regime change operation. First, North Korea's "winning in peacetime" strategy would be characterized by nuclear threats to establish a stability/instability paradox in East Asia whereby Pyongyang could reshape regional relationships to its liking. If that strategy failed and if the United States or an ally initiated military strikes against Pyongyang, then a "winning a limited war through nuclear blackmail and brinksmanship" strategy would witness the limited introduction of its nuclear queen to frighten Washington from taking any further military action, solidify the Kim regime's hold on power, and restore or improve upon the *status quo ante*. The worst-case scenario strategy of "Winning a Total War begun by the United States" would witness Pyongyang escalate its use of nuclear force against the U.S. homeland and its allies, even if it led to North Korea's destruction. Indeed, with this last strategy, a humiliating capitulation to the United States would not be regarded as a permissible option. Instead, honor would require embracing nuclear death while simultaneously inflicting horrific damage on the American aggressors.[36]

In this vein, Roberts relates that U.S. administrations since the Cold War's end have concluded that the United States should avoid entering into new mutual deterrence relationships with NWS adversaries because they would leave Washington too vulnerable to nuclear blackmail, reduce the credibility of its nuclear threats below the level of armed conflict, and fail to properly reassure allies that rely on U.S. extended deterrence.[37] Rather, it should continue to develop a comprehensive effort to tailor and strengthen deterrence for each region of concern. This would involve maintaining current alliances and building new partnerships with other states on matters of shared interest. Additionally, the United States should deploy conventional forces capable of effective and quick defensive operations and long-strike capabilities, deploy ballistic missile defenses to protect allies and forward-deployed U.S. forces, improve capabilities to address cyberattacks from regional challengers, and, finally, express unambiguously the United States' intent to defend its interests and those of its allies.[38] Only such a comprehensive approach would enable the United States to effectively counter the use of a challenger's "atomic queens."

The United States' stance against new relationships of mutual vulnerability with new nuclear states recalls the Cold War debate on the morality of MAD between the United States and the former Soviet Union (see chapter 2). Advocates of MAD believed that mutual vulnerability to nuclear threats was a necessary condition for U.S.-Soviet strategic stability and that, from a moral consequentialist viewpoint, the desired end of war prevention was enough to excuse or justify the imposition of counter-city nuclear threats. Indeed, some security theorists had argued that the mutual vulnerability produced by nuclear deterrence was consistent with a common security conception,[39] which is discussed later in the chapter. Critics believed that the mutual vulnerability of civilians of adversary nuclear states was contrary to the *jus in bello* noncombatant immunity principle, which marked an important limit on the self-defense rights of states. The political and moral question centered on the feasibility of purely counterforce nuclear defense and deterrence, which eliminated mutual vulnerability as a necessary condition of international stability and security. Critics believed that the technological recalcitrance of nuclear weapons (i.e., the impossibility of nuclear weapons' discriminate use) and the problems of escalation control would likely subvert an effective counterforce policy aimed at restoring deterrence after nuclear first use.[40] However, these criticisms have not influenced the way U.S. administrations after the Cold War have thought about this issue. The U.S. stance against mutual deterrence relationships now rests on a collective security imperative to ensure the proper imbalance of power so that "rogue state" challengers are effectively dissuaded from inducing instability.[41] This stance implies that, if mutual vulnerability was a moral virtue during the Cold War, it was no longer in post–Cold War contexts (even if some degree of mutual vulnerability with Russia and China remains unavoidable).

Roberts's prescriptions also recall the Cold War arguments by antinuclear advocates on the preferability of conventional over nuclear deterrence. The later part of chapter 2 related Steven Lee's argument that nuclear deterrence is morally justifiable only if conventional deterrence could not provide an adequate alternative. For the Obama administration, the security challenges of the early to mid-2010s did not permit a wholesale switch from nuclear to conventional deterrence, even though it sought to reduce the role of nuclear weapons in U.S. security policy. Roberts notes that ballistic missile defense as a kind of conventional deterrence can effectively signal assurance to allies and, in some cases, positively affect nuclear deterrence postures. However, it remains that "missile defense is a complement to nuclear deterrence and not a substitute."[42] Similarly, he argues that conventional rapid-strike capabilities, such as the "shock and awe" campaigns which the United States executed against Saddam's Iraq, will not affect adversary calculations as would nuclear capabilities since they do not provoke in the enemy fears of a "sudden and complete loss" of vital military and political assets. Last, conventional military supremacy is not likely to enhance deterrence since the U.S. military has endured budget cuts and endured fatigue in the course of two decades of constant warfare.[43] If Roberts' analysis is correct, then at least one line of Cold War opposition to nuclear deterrence appears to not pass muster.

The most recent phase of the evolution of U.S. nuclear defense and deterrence policies followed Donald J. Trump's election to the U.S. presidency in November 2016. It is widely argued that the first three years of the Trump administration were turbulent in both domestic and foreign affairs, especially as Trump raised the prospect of "fire and fury" against North Korea, insulted European allies, criticized China, and (oddly) praised Vladimir Putin.[44] Regarding the North Korean nuclear crisis in particular, Trump's initial posture of hostility turned to conciliation during the 2018 Singapore Summit, where it seemed that Kim Jong-un had committed to denuclearization even though he made clear that it must follow the formal end of the Korean War and withdrawal of U.S. nuclear capabilities from the region. This difference of viewpoints between Washington and Pyongyang was enough to demonstrate that the North Korean nuclear challenge was not going away anytime soon.[45]

In conjunction with these developments, the administration's 2018 Nuclear Posture Review (NPR) made clear that "America confronts an international security situation that is more complex and demanding than any since the end of the Cold War."[46] Specifically, it highlighted the "return of great power competition" as Russia and China sought to "substantially revise the post–Cold War international order and norms of behavior."[47] It also contended that Moscow regarded NATO and the United States as its principal strategic enemies. Moreover, it noted that Moscow had begun to emphasize the utility of nuclear-first-use with low-yield weapons to paralyze U.S.-NATO decision making and thereby increase direct control or indirect influence over neighboring countries.[48] Regarding China, the 2018 NPR noted that Beijing had committed to nuclear modernization and to the continued pursuit of regional dominance.[49] Finally, it marked the continuing geopolitical and nuclear challenges posed by North Korea and Iran.[50] For these reasons, it argued that the United States should modernize its nuclear forces to effectively field a tailored nuclear deterrent instead of the "one-size-fits-all" strategy employed during the Cold War.[51]

The upshot is that the Trump administration's focus shifted from Obama's emphasis on the prevention of nuclear terrorism and new nuclear proliferation to the more traditional concern of addressing great power security competition. This shift of emphasis might explain the bipartisan support in the U.S. Congress for nuclear modernization.[52] As Keith Payne put it,

[we must understand] the variability among opponents and the corresponding need to tailor deterrence strategies . . . [as well as] the need for considerable flexibility and diversity in our deterrence strategies and capabilities. The more dynamic and uncertain the threat environment, the more important is the flexibility of our deterrence planning and diversity of our threat options.[53]

This shift in focus, combined with the Trump administration's abandonment of a foreign policy consistent with the norms of a liberal world order and a significant sizing down of the U.S. State Department, suggests an additional shift toward a greater reliance on coercive diplomacy backed by nuclear force.[54]

If the foregoing sketch of the evolution of U.S. nuclear policy is correct, then two conclusions are suggested for the book's task of rethinking nuclear ethics for the twenty-first century. One is that the "fundamental structural circumstances" of the post–Cold War international order are not, in one sense, significantly different from those of the Cold War era.[55] This is to say, the fundamental dynamics of interstate security competition remain in place despite the shifts from bipolarity to unipolarity to multipolarity over the past three or so decades. Moreover, the United States and Russia possess over 90 percent of the world's nuclear weapons and thus still dominate the field of nuclear defense and deterrence dynamics.[56] On the other hand, these shifts in polarity might be enough to motivate some revision of past nuclear ethical codes. This latter question will be deferred to later chapters, but it is time to conclude the historical sketch with a few reflections on the Trump administration's nuclear policy.

Most importantly, the Trump administration ought not to blithely assume immunity from the increasing risks of deterrence failure. According to Ulrich Kuhn, such risks are most likely to arise from this administration's lack of appreciation of the asymmetries of preference intensity regarding competing states' interests.[57] For instance, Moscow is clearly more interested in establishing buffer zones against NATO than is Washington in committing all required resources to continue NATO expansion, rollback Russian forces from Ukraine, adequately resolve the anxieties of NATO's easternmost members, and so forth. This point is evidenced especially in President Trump's confrontational relationship with NATO and his affinity for Russian president Vladimir Putin.[58] Kuhn's analysis holds also for North Korea: Pyongyang's interests in regime security and retaining a nuclear deterrent capability are significantly more intense than the U.S. interest in doing everything necessary to compel Pyongyang's nuclear rollback or to overthrow the Kim regime.[59]

If Kuhn's analysis is correct, then there is a moral imperative to adopt a different strategy than what the 2018 NPR recommends and what the Trump administration has done so far to address the geopolitical conflicts of interest among the nuclear-armed powers. To that end, Ken Booth and Nicholas J. Wheeler's book-length analysis of competing security logics, which included an examination of Deutsch's security community and Palme's common security approaches, cogently argues that "the ultimate insurance against war . . . lies in political community, not nuclear threats. Predictable peace comes through norms, institutions, laws, multilevel social interaction, complex interdependence, trust-affirming commitments, and the promotion of cosmopolitan sensibilities."[60] An application of Booth and Wheeler's analysis to the field of nuclear ethics recalls one question over which Cold War–era nuclear ethicists disagreed: namely, is the state or are individual persons (and thus humanity itself) the ultimate referent of security understood as a moral value? Answers that privilege the former suggest that the security of the individual is ultimately dependent upon state security. On this view, policymakers can then claim moral justification to use nuclear threats to prevent warfare. Answers that privilege the latter assume that the right of state security is derivative on the right of individual security. Such answers have tended to reject nuclear defense or deterrence as morally acceptable ways of ensuring state security. And yet, even during the Cold War, it was evident that the fate of states and individuals (and humanity) was intertwined—that is, their security was interdependent.[61] Booth and Wheeler's argument thus raises a question about the possibilities and limits of Deutsch and Palme's innovative common security conception in recommending a new anchoring principle for nuclear ethics.

A COMMON SECURITY FRAMEWORK

This chapter's introduction contended that the political logic of common security should replace the predominant conceptions of national, collective, and human security around which the international nuclear order in its complexity has been established. The key conceptual difference of common security is expressed by the phrase "security with" adversaries rather than "security against" them. As Figures 3.1 and 3.2 illustrate, the "security against" conception is the fundamental assumption of both national security and human security frameworks.

In the national security framework, each state is secured "against" aggression by means of unilateral military deterrence or defense postures. Similarly, the human security framework is advanced in the context of oppressed minority populations that seek "security against" fundamental rights abuses at the hands of their governments or their government's proxies. In some cases, this means the willingness of a neighboring country to provide sanctuary to refugees. In other cases, it means military action under the rubric of humanitarian intervention. As Figure 3.3 illustrates, even some forms of cooperative security—for example, collective security—assume the "security against" conception in relation to "rogue" states or rival alliances, as was evident in the Cold War between NATO and the Warsaw Pact.

By contrast, the common security framework proposes a "security with" formulation between or among state adversaries, especially those with the capacity to wage wars of extermination. As Palme remarked during the early 1980s, when the superpowers restarted the nuclear arms race,

> There can be no hope of victory in a nuclear war. The two sides would be united in suffering and destruction. They can only survive together. They must achieve security not against the adversary but with him. International security must rest on a commitment to joint survival rather than on a threat of mutual destruction.[62]

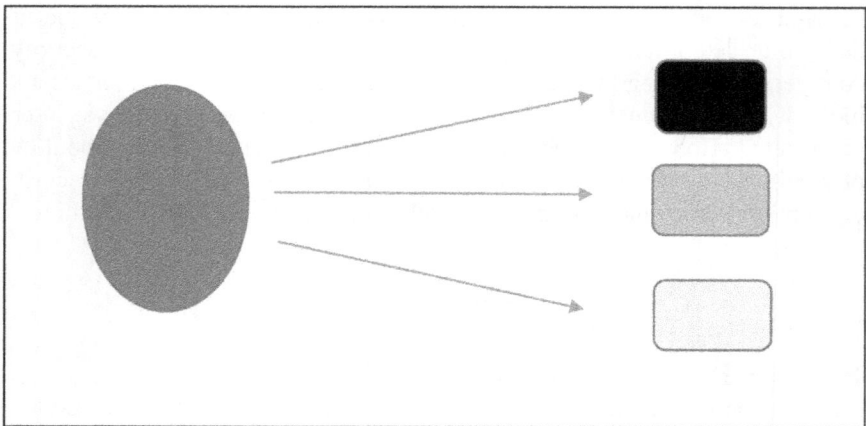

Figure 3.1.   National Security Framework
Key: Oval—the homeland; rectangles—the rivals of one's homeland; arrows—homeland's security measures against state rivals.

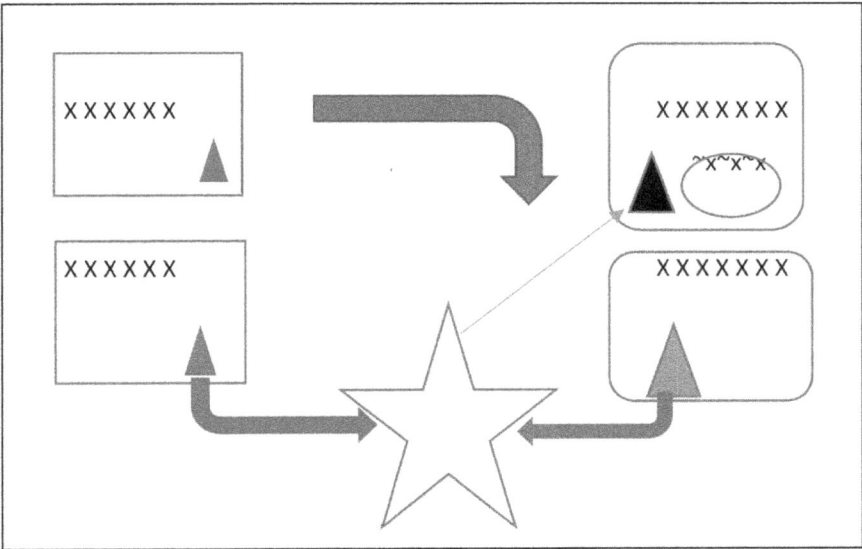

Figure 3.2. Human Security Framework
Key: rectangles—liberal states; rounded rectangles—illiberal states; X's—citizens of states; ~x—citizens (at risk of) suffering human rights violations; gray triangles—governments respectful of fundamental human rights; black triangle—government violating fundamental human rights; five-pointed star—international organization; thick arrows—participation in international organization; thin arrow—security measures against governments that violate fundamental human rights.

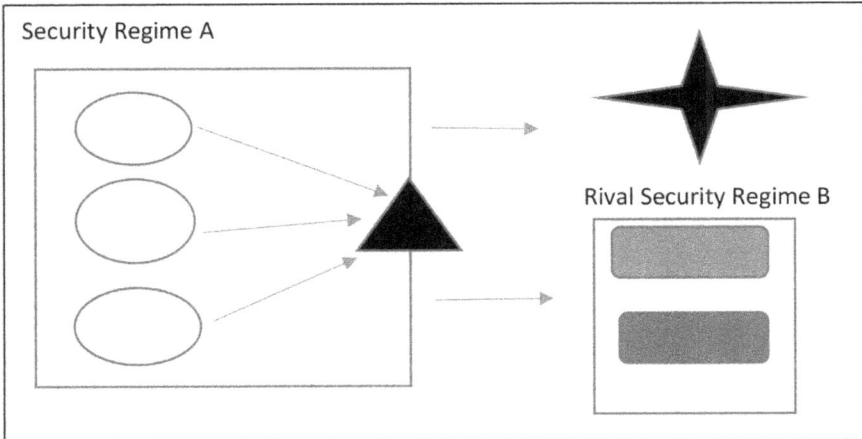

Figure 3.3. Collective Security Framework
Key: Ovals and rectangles—states in a collective security regime; squares—collective security regime; triangle—rogue state within a collective security regime; four-pointed star—rival unaligned state; arrows—security regime A's measures against rogue actors and rivals.

For Palme, a nuclear war would be a futile endeavor for each superpower, even if one or both happened to survive and claim victory. Palme doubted strongly that neither superpower nor its neighbors could avoid catastrophic damage following any limited nuclear conflict, and that any mutual escalation of nuclear conflict would eventually lead to their mutual annihilation. In the end, the nuclear security dilemma forces a recognition and acceptance, on pain of mutual annihilation, that survival is a common interest and value among all nuclear-armed enemies, their allies, and their peoples. And the realization of that common security interest requires a specific kind of security cooperation that is not found in other cooperative security approaches (e.g., collective security).[63]

Karl Deutsch on Security Communities

Karl Deutsch's account of security communities preceded Palme's account of common security by approximately twenty-five years.[64] Deutsch defined a "security community" as a group of distinct political communities that integrated by a set of institutional practices emphasizing the peaceful resolution of conflicts of interest.[65] He distinguished between amalgamated security communities produced by the merger of two or more political communities into a formal union (i.e., the United States, the United Kingdom) and pluralistic security communities produced by looser confederal arrangements in which each member state maintained its sovereignty (i.e., the EU). His empirical research showed that, since the late eighteenth century, the construction of both kinds of security communities among North Atlantic and European states was not infrequent. Additionally, he discovered seven instances of serious failure of amalgamated security communities in comparison to the one instance of a failed pluralistic security community (i.e., Austria and Prussia in the nineteenth-century German Confederation).[66]

From this research, Deutsch concluded that successful integration requires several prerequisite conditions. For pluralistic security communities, success depends on actors' shared commitment to specific modes of political decision making, the capacity of each for effective and mutual responsiveness to each other's needs, and the capacity for mutual predictability of behavior, especially on major policies.[67] By contrast, he found that disintegration is more likely if one or more of four developments arise, each of which undermines the maintenance of shared values, mutual responsiveness, and mutual predictability of behavior. First, disintegration is likely to occur when excessive military burdens are placed by one member on fellow members. An inequality of burden sharing demonstrates insensitivity and selfishness, which is corrosive for the community's shared security interest. Moreover, inequality of burden sharing is unfair and unjust unless it can be adequately defended to all relevant others' satisfaction (see chapter 6 on the role of fairness in justice). Deutsch cited as an example the eighteenth-century Imperial British imposition of excessive military burdens on their American colonies in the aftermath of the Seven Years War, which began the process of disintegration culminating in the 1774–1783 American Revolutionary War.[68]

Second, disintegration is more likely when segments of a community's population that had not been politically active rise to make their voices heard. Not only does their new political activity signal a lack of satisfaction with the *status quo*, but it identifies the extent to which values are contested within what had appeared to be a unified community. Deutsch cites the example of the anti-slavery movement in the United States between 1830 and 1860, which, as it challenged a pro-slavery *status quo*, contributed to the Civil War between 1861 and 1865. Third, disintegration is likely when there is an unanticipated or unwelcome increase in ethnic or linguistic differentiation within a community. Unless such differences can find accommodation, they directly weaken a community's capacities of interpersonal communication and the ability to maintain a set of shared fundamental values. A recent example is suggested by various European nationalist and populist political parties whose opposition to African and Middle Eastern immigration seems driven by the desire to maintain white racial dominance.[69] Finally, disintegration is likely when there is an unacceptable delay in anticipated political, economic, or social reforms within a community. The impatient demand for such reforms constitutes a call by one member of the community for some mode of *security against* the wider community. Prompt implementation of reforms, or those that did not suffer unacceptable delays, would not seem to threaten the core "security with" dynamic among community members. However, unacceptable delays signal that each necessary condition for the success of a security community is not adequately in place or has disappeared.

Some commentators seem to associate in a rather limited way conceptions of cooperative security with that of collective security. For instance, the Organization of Security and Cooperation in Europe (OSCE) applies a "comprehensive concept of security" to its members such that military, economic, and human aspects of security are interwoven and thus mutually reinforcing.[70] Put in slightly different terms, each OSCE member "has an equal right to security" based on the notion that "security is indivisible and that the security of each . . . is inseparably linked to the security of all."[71] Interestingly, Wolfgang Zellner finds that effective European cooperative security is difficult to effectively implement and maintain because some states are exposed to specific risks and threats to which other states are not and because each state continues to place its own interests first, even if it means engaging in zero-sum strategies.[72] Zellner's observations thus recall Deutsch's second condition of integration failure, suggesting that the OSCE security conception has not enabled the realization of indivisible security for all.

Holger Molder's findings for the Baltic Sea states on questions of cooperative security echo Zellner's analysis.[73] Molder argues that Baltic Sea states sought in the 1990s to form a "Kantian security culture" hallmarked by security cooperation and mutual interdependence. Even so, these states became ensnared in a series of "cooperative security dilemmas" around the issues of integration and identity harmonization. These dilemmas were encouraged by the failure of a Kantian security culture to fully permeate Europe and the Baltic states. Accordingly, Molder finds the emergence of several security paradoxes including intensified competition and identity-related clashes between Western and Eastern EU states, the latter of which border Russia and which have increasingly embraced illiberal dynamics. On the basis of Molder's analysis, it seems that even if the Baltic states were to effectively resolve their integration dilemmas, they would still face identity dilemmas with Russia, as well as among states whose populist parties are experiencing a significant resurgence.[74]

The foregoing helps to show that not all cooperative security arrangements are common security arrangements as defined by Palme's "security with" conception, but all common security arrangements are cooperative security arrangements. It might be that augmenting collective security is important or necessary to provide the opportunity for common security arrangements to work. However, cooperative security as collective security has not and will not be enough to resolve the geopolitical conflicts of interest between nuclear-armed antagonists, and as such it is not a sufficiently compelling security conception upon which to anchor a twenty-first-century nuclear ethics.

Olof Palme on Common Security

Almost three decades after Deutsch's work on security communities, Palme's commission authored a report that linked the concept of common security to the survival of each state and people in the nuclear age.[75] As leader of Sweden's Social Democratic Party and as Sweden's prime minister from 1969 to 1976 (and then again from 1982 to 1986 after the commission's report had been published), Palme had acquired a reputation as a staunch critic of the superpowers' foreign policies that he described as overly militaristic and neo-imperialistic.[76] He did not believe that the United States or Soviet Union could be persuaded to embrace nuclear disarmament if their geopolitical competition continued. Thus, the security of Sweden, of any other moderately powerful state similarly situated, of the superpowers themselves, and of humanity itself required rethinking the conditions of regional and international security generally. Accordingly, Palme argued,

> In the nuclear age, no nation can achieve absolute security through military superiority. No nation can defend itself effectively against a nuclear attack. No matter how many nuclear weapons a nation acquires, it will always remain vulnerable to

a nuclear attack. And thus, its people will ultimately remain insecure. . . . [Indeed, security] can thus not be achieved through unilateral measures—there is no such thing as a modern *Pax Romana*. Security must instead by achieved through cooperative efforts. Even political and ideological opponents must work together to avoid nuclear war. They can survive only together. They would be united in their destruction. . . . Security in the nuclear age means common security.[77]

Palme implies that, prior to the nuclear age, national security could be realized by national defense traditionally conceived. However, nuclear-power-based aggression exceeds the capabilities of any existing conventional-weapons-based national defense capability. And, if states are defenseless against such nuclear attacks, so are their peoples.[78] Accordingly, the common insecurity of peoples and states produced by the constant threat of nuclear annihilation establishes the groundwork for a new security conception—that "security in the nuclear age means common security." Figure 3.4 attempts to capture Palme's complex conception.

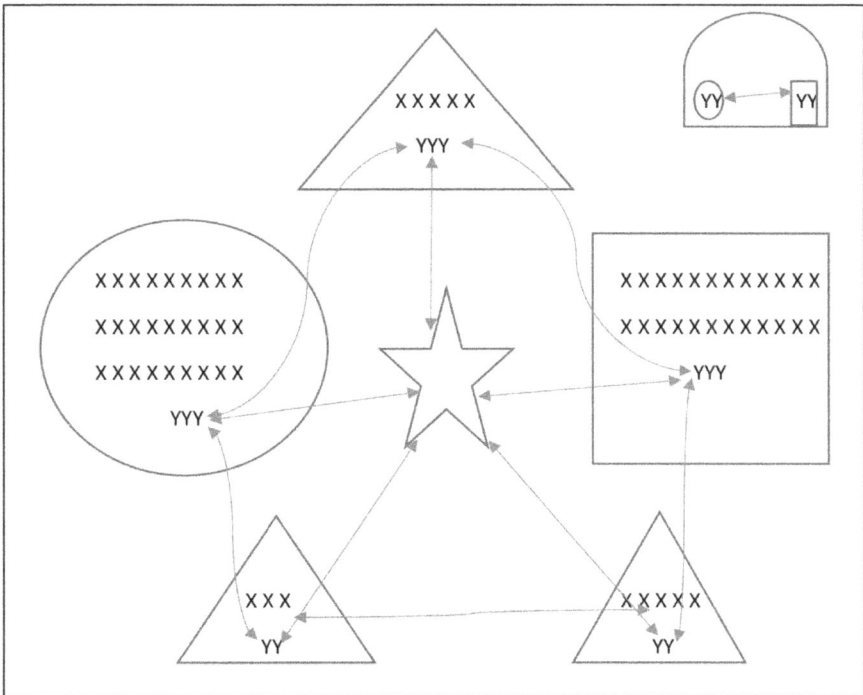

Figure 3.4. Common Security Framework
Key: circle—the homeland; square—state adversary to the homeland; triangles—third-party states to the adversarial relationship between homeland and adversary; five-pointed star—international organization; Xs—citizens of states; Ys—state governments; half-dome—face-to-face diplomacy venue for homeland and adversary government leaders; bidirectional arrows—security "with" interactions.

Palme conceived of common security in political and moral terms. By linking the fate of humanity to the fate of states in which humanity's peoples live, he strongly suggested that the distinct and competing moral imperatives of state and human security are reconciled through the common security framework. This reconciliation is conceivable by analyzing "the components of that structure of common security" for the early-1980s nuclear threat environment.[79] First, Palme believed that one essential component of that structure was military parity between the superpowers and their respective alliance blocs. The United States and Soviet Union had achieved rough nuclear parity by the early 1980s, and he believed that achievement made possible a negotiated parity between NATO and Warsaw Pact conventional forces.[80] Once conventional parity is reached, then further negotiated settlements would replace nuclear arms racing as an enduring means of maintaining security. Accordingly, Palme anticipated the successful conclusion of a comprehensive nuclear-test-ban treaty, a chemical weapon ban treaty, and the achievement of universal membership in the NPT.[81] As each new agreement was concluded, Palme believed that the anti nuclear and enhanced anti-WMD norms would be strengthened and thus contribute to the increased security of all states and peoples.

Second, Palme emphasized the necessity of regional common security orders for the construction of global common security. He was concerned about the persistence of wars among states in Latin America (e.g., El Salvador), the Middle East (e.g., Israel-Palestine, Iraq vs. Iran), and South Asia (e.g., India and Pakistan over Kashmir), which would continue to provoke global great power rivalries.[82] In one instance, he argued,

> The Palestinian question, to mention one example, cannot be solved by the destruction of the PLO (the Palestine Liberation Organization). Desolation is not peace even of [sic] you call it that. . . . Security in this region must be common security.[83]

Palme's remarks emphasize that the security of Israel is not a function of the elimination of the PLO. "Desolation is not peace" or security, even though each adversary might stubbornly insist that the other's elimination or sublimation is the only "realistic" option. Rather, Israel's security cannot be divorced from Palestinian security. Two additional propositions followed Palme's contention. One was that common security could be applied across various levels of analysis: that is, it was essential for intrastate as well as interstate conflicts. The other was that the greater powers' security interests are often implicated in those of the lesser powers.[84] Hence, if regional conflicts could induce great power warfare, then regional common security arrangements could facilitate the same among great powers.

Finally, Palme emphasized the necessity of international institutions and law for an enduring common security order. Given the United Nations' mixed record of providing for regional security, Palme urged that it be "used in a more determined way" and that "its security role must be strengthened,"[85] such as more effective Security Council deployments of peacekeeping forces in troubled regions and states to better facilitate successful negotiated settlements between combatants. It also meant a strengthened secretariat that, in concert with the permanent members of the Security Council, would more effectively prevent regional conflicts from starting.[86] Of course, the strengthening of the United Nations could not be accomplished without a corresponding commitment by the UNSC permanent members to the "universal adherence to the rules of international law."[87] Such a commitment would require each of the UNSC permanent members to abstain from excusing themselves from the requirements that they placed upon other states in the name of national security.[88]

In the end, Palme's common security conception amounted to a theoretical and practical innovation in International Security Studies that Barry Buzan and Lene Hansen believed constituted "the single most expansive concept of the 1980s."[89] Buzan and Hansen concurred with Palme that the main threats to international security

> came not from individual states but from global problems shared by the entire international community: nuclear war, the heavy economic burden of militarism and war, disparities in living standards within and among nations, and global environmental degradation.[90]

Deutsch and Palme's accounts did not reduce state insecurity to exogenous threats of aggression, nor did they privilege individual persons or humanity in general as the principal referent of security. Rather, as Figure 3.4 illustrates, the "global problems shared by the entire international community" required a focal point of analysis and policy through which the security interests of diverse kinds of actors (i.e., individuals, peoples, and states) might be reconciled under the formulation of each achieving "security with" the other(s).

In recent history, some elements of the common security framework have been instituted among North Atlantic and European states with varying degrees of success. Germany and France, which had been historic rivals for centuries until the end of World War II, redefined their relationship around common values, interests, and institutions and, as a result, have abandoned military means of resolving their mutual conflicts. The end of the Cold War prompted the United States and NATO to attempt a similar redefinition of relationship with the Russian Federation. In 1991, the North Atlantic Cooperation Council was established to facilitate security cooperation and peace between NATO and former Warsaw Pact states. In 1994, the Partnership for

Peace program was established between NATO and the Russian Federation, with the aim of including Moscow in discussions and decisions regarding NATO policy.[91] Unfortunately, these efforts stalled in the wake of NATO expansion and the Bush 43's decision to replace deterrence with missile defense.[92] As great power relationships have continued to fracture since the mid-point of the Obama administration, it has become evident that the likelihood of inducing a common security dynamic among the great powers—and specifically the United States, Russia, China—is diminishing. Its diminishment is captured by the Bulletin of the Atomic Scientists' determination in 2017 and 2018 that the Doomsday Clock is once again set at "two minutes to midnight" (see chapter 1, introduction). This stark reality suggests the theoretical and practical limits of common security.

The Limits of Common Security: Ontological Security and Ideological Fundamentalism

Before considering common security as an anchoring principle of a new twenty-first-century nuclear ethics, it is important to consider its limits. This subsection focuses on the limits imposed by ontological security-seeking by states as conditioned by ideological fundamentalism.[93]

For mainstream IR realists, states are rational security-seekers that avoid taking any overly risky actions in pursuit of an interest not related to ensuring physical survival. In contrast, some scholars argue that states have taken "irrational" risks to protect and preserve a consistent set of national self-identity conceptions that define the "soul" of a country.[94] This kind of security pursuit is consistent with the conception of "ontological security." Brent Steele offers the following contrast between "rational" and ontological security-seeking:

> While physical security is (obviously) important to states, ontological security is more important because its fulfillment affirms a state's self-identity (i.e., it affirms not only its physical existence but primarily how a state sees itself and secondarily how it wants to be seen by others). Nation-states seek ontological security because they want to maintain consistent self-concepts, and the "Self" of states is constituted and maintained through a narrative which gives life to routinized foreign policy actions. Those routines can be disrupted when a state realizes that its narratived actions no longer reflect or are reflected by how it sees itself. When this sense of self-identity is dislocated an actor will seek to re-establish routines that can, once again, consistently maintain self-identity.[95]

Steele's remarks indicate that ontological security-seeking behavior is concerned with the essence of a state's "soul" as constituted in its national narratives over time. At the domestic and international levels, Steele finds that the struggle for political identity is as or more intense than that to maintain territorial integrity. Instead of an all-consuming fear of the loss of physical existence, states are more fearful of experiencing shame—that is, a deep anxiety over the inability to reconcile their experienced reality with their national identity narratives.[96] An intensely nationalistic identity conception is strongly associated with "security against" narratives, and its use in domestic political discourse impedes efforts to otherwise establish a common security relationship with historic enemies. Table 3.3 relates the main distinctions in Steele's account between the traditional national security and the ontological security frameworks.[97]

In support of his theory, Steele examined Belgium's 1914 decision to resist the German ultimatum that had demanded for its army unfettered access through the former's territory to invade France early in World War I.[98] Steele argues that Belgium's decision is inexplicable from a mainstream IR realist perspective. Germany had overwhelming power and Belgian defiance of this ultimatum would lead to the latter's certain defeat. Even so, Belgium clearly expressed its defiance and, in the end, it suffered 30,000 deaths, thousands more injured, over one million displaced, and approximately 115,000 deported to German labor camps. Despite these horrific losses, the official Belgian discourse (as well as that of its French and British allies) consistently elevated Belgian honor as a neutral European state over its physical security.[99] As Jack Donnelly argued in relation to the IR realist paradigm case of the Athenians and Melians in the Peloponnesian War,

> Honor . . . demands that one die fighting rather than submit. . . . The Athenians forcefully and effectively lay out the Melian interest in capitulation. The Melians, however, are willing to die rather than live with the shame of

Table 3.3 Contrasting Traditional and Ontological Security Frameworks

	National Security (Traditional)	Ontological Security
Security as	Survival	Being
Structured by	Power distribution	Behavioral routines, self-description
Challenged by	Fear (in the face of external state adversaries)	Anxiety (uncomfortable disconnect with the national self; cognitive dissonance)
Failure of seen by	Suffering of physical harm or loss	Suffering of shame independent of physical harm or loss
Measuring outcomes by	Gains or losses in material capabilities; deaths, injuries, damage	Consistency or inconsistency of identity narrative with state behavior; pride or remorse; gains or losses in reputation
Structural change	Changes in the distribution of power	Changes in state routines, self-identity narratives

submission. . . . The Melians begin with an appeal to justice. They understand
their interest and come to grips with their fear. But, in the end, the Melians die
for their honor.[100]

Steele quotes Donnelly's analysis to demonstrate that the Belgian government put its national self-conception over physical survival because to do
otherwise (i.e., to capitulate to the German ultimatum) would produce an
intolerable degree of shame and remorse.

In a contrasting study of ontological security-seeking by NATO member
states during the 1999 Kosovo operation against Serbia and its ethnic cleansing of Albanian Serbs, Steele demonstrates that the primary concern of the
alliance's great power actors was not fundamentally about the Albanian
Serbs' welfare. Rather, it was the resolution of each power's shame over
their respective foreign policy failures.[101] To cite only two examples, the
United States sought to overcome the shame over its inaction during the
Rwandan genocide while claiming to be the world's leader on human rights.
For its part, Germany sought to overcome the guilt of its Nazi past in order to
become a "normal" state again, especially as the United States had put more
demands on Berlin for increased NATO involvement.

These case studies suggest that states' pursuit of ontological security does
not necessarily require a strong preference for "security against" over "security with" approaches. Instead, this choice depends significantly on the content of a given state's identity conception.[102] For instance, the key adversarial
relationships (i.e., Belgium vs. Germany; NATO vs. Serbia) were inextricably defined by "security against" approaches, even if a measure of "security
with" was established among allied states (e.g., Belgium and France, NATO
member states).[103] These comparisons suggest that "security against" under-
girds both mainstream national security and ontological security frameworks
in spite of their other core differences. And if the ontological security framework offers a better explanation of security-seeking behavior by states, then
we need an account that explains how it can limit the realization of common
security. The next few paragraphs explore these limits in relation to Booth
and Wheeler's analysis of ideological fundamentalism.

Booth and Wheeler define ideological fundamentalism as a state's disposition to assign "enemy" status to a state adversary "because of what the other
is—its political identity—rather than how it actually behaves."[104] A state's
depiction of the other's identity in terms of beliefs, values, narratives, and
culture provides a simple measure of assigning friend or foe status than does
their behavior alone, much of which is fraught with ambiguous meanings.
And if the other state's ideology is opposite to one's own, it is relatively simple to depict them as an enemy posing an existential threat.[105] Thus, whereas
a state pursuing ontological security attempts to stabilize and preserve a

favored national identity from threats, ideological fundamentalists seek to threaten or destabilize their adversary state's identity conception. Ontological security and ideological fundamentalism are thus mutually implicated.

As a result, a state's only remaining challenge is to choose how best to "secure against" its ideological foes. Since negotiations leading to rational compromises can be perceived as an unwanted appeasement by ideologues, it often happens that the only permissible response for an ideologically based foreign policy is coercive diplomacy or military force.[106] The practical implications of ideological fundamentalism in U.S. domestic and foreign policy have been evident in its resolute opposition to Marxism and Soviet/Chinese communism from the late 1910s through the early 1990s: for example, the Red Scare, McCarthyism, loyalty oaths, and Cold War–era nuclear deterrence.[107]

Two other remarks about ideological fundamentalism in relation to ontological security are important to make. First, an unwavering ideological opposition to national differences is a necessary condition of reaffirming or policing nationalist identity conceptions and practices among one's own citizenry. According to David Campbell, it is impossible to construct a national identity conception in a final or unalterable way, for the never-ending changes in domestic and international political dynamics always mean that there is something new for identity conceptions to address and articulate. The construction of national identity is always a work in progress, and therefore it is always insecure. Campbell's account demonstrates that Americans used their ideological depiction of the Soviet Union for reassurance that they were truly identified with the causes of individual freedom and dignity, representative government, the rule of law, human rights, free expression of religion, and so forth. As Campbell demonstrates, through several publications and official policy pronouncements (e.g., NSC-68, Reagan's Evil Empire Speech), U.S. political and cultural leaders reinforced the narratives of American self-identity, named and shamed those Americans who sought to express dissent, and legitimated actions that might seem otherwise contrary to American values in the name of containing or defeating Soviet adventurism.[108] As hinted at earlier, ideological fundamentalism became the chosen means of ontological security for the United States.

Second, in the foreign policy realm, ideological fundamentalism pushes state actors to resist détente relations with adversaries in the name of never compromising with "evil." This point is illustrated by the failure of U.S.–Soviet détente during the 1970s, which the Nixon administration pragmatically introduced to achieve mutual stability as both superpowers reached nuclear parity.[109] Détente collapsed not long afterward, during the Carter administration, as hawkish critics reasserted an ideologically charged opposition against the continuation of U.S.-Soviet nuclear arms control and eventually won the White House with the election of Ronald Reagan in 1980. These critics argued that the Soviets would not accept parity but instead seek nuclear superiority in its bid to prevail against the West and its way of life. And even though there was credible evidence suggesting that the Soviets had accepted détente and nuclear parity, the hawks' ideological commitments constrained them to interpret this evidence as a Soviet strategic ploy to lure the Americans to sleep and establish nuclear superiority before Washington could effectively respond.[110]

The implications of ideological fundamentalism as a principal mode of seeking ontological security seem altogether negative for the possibilities of constructing common security relationships between historic ideological foes. For one, it seems that ideological fundamentalist approaches cannot conceive of a political or social space where "good" and "evil" actors can identify a common set of core values, beliefs, or identity conceptions. Any accord between ideological opponents would inevitably reflect "security against" postures and would always be vulnerable to subversion by one or both parties. Palme's urgent call to reconfigure superpower relations around the common security conception thus sought to displace the superpowers' ideological fundamentalisms. And although the Cold War ended a decade later with fledgling attempts at closer security cooperation, the United States and Russia remained fixed in their respective "security against" postures as NATO expanded and Russia felt increasingly contained.

COMMON SECURITY: AN ALTERNATIVE ANCHORING CONCEPT FOR NUCLEAR ETHICS

The preceding sections set the stage for the chapter's final task of proposing the common security conception as an alternative anchoring moral principle for nuclear ethics in the twenty-first century. It has been suggested that Palme's conception of "security with" is a middle-ground position reconciling the national and human security frameworks. If this is true, then a common security imperative does not require the trade-off of national security for human security, which is a principal concern of nuclear defense and deterrence advocates. Alternately, it does not require the trade-off human security for national security, which is a principal concern for nuclear defense and deterrence opponents. Rather, it redefines the political and moral relationship between or among state adversaries such that national and human security imperatives are satisfied by the mutual shift in practices by states and peoples from "security against" to "security with." Hence, common security might be understood as a bridge principle of nuclear ethics capable of reconciling the national and human security imperatives.

As I see it, a formulation of common security as an anchoring and bridging concept of nuclear ethics is best accomplished by way of Kantian theory. According to Mary Gregor, Kant used his 1797 *Metaphysics of Morals* in part to articulate systematically the grounds on which his 1795 essay *On Perpetual Peace* (PP) rested, the grounds of which stated that "morally practical reason pronounces in us its irresistible veto: there is to be no war, neither between you and me in a state of nature, nor between us as states."[111] For Kant, this moral imperative against warfare could not be overridden and it must ultimately override all other competing political interests that would justify warfare among states. The authority of Kant's anti-war imperative is derived from the authority of practical reason, which regulates all modes of possible social activity. Accordingly, Kant's aim in PP was to apply this anti-war imperative to international politics by enumerating a series of preliminary and definitive articles that specified the necessary practical conditions of perpetual peace. In the end, Kant's pacific federation might be seen as a precursor of a Deutschian security community or, as Dora Ion puts it, of a "cosmopolitan communitarian" way of being.[112]

Kant's first preliminary article contends that peace treaties are not the same as armistices or truces, which can provide only for a temporary cessation of hostilities as each adversary continues to struggle to gain strategic advantage. Rather, for Kant a peace treaty "annihilates" all causes for future wars among state adversaries.[113] His use of the term "annihilates" refers to an absolute and fundamental change in each state's identity, interests, and status in relation to its adversaries: that is, each state's ontological security. Prior to the peace treaty process, state adversaries regard their mutually conflicting identities, interests, and statuses as matters of unchanging social and political reality. However, the process of constructing peace (as opposed to truce) arrangements means that, at some moment during the conflict, they undertake a mutual commitment to the moral imperative of common security. In the process, each undertakes with the other(s) a series of steps that deconstruct their ideological antagonisms that had motivated their existential fears. By the end of the process, the identity and interest conceptions of each have been mutually reconstituted such that each no longer regards the other as an enemy. The abolition of the mutual enmity relation enables their mutual realization of "security with." Understood this way, the peace treaty codifies or encodes a crucial element of their common security bond, which might culminate in their joint entrance into a pacific federation of states.

None of the foregoing implies that Kant believed the process of perpetual peace is easy or quick. For example, PP's third preliminary article contends that "standing armies shall in time be abolished altogether."[114] Kant understood that, prior to the conclusion of a genuine peace process, states would always prepare for offensive or defensive military operations. Such preparations directly and forcefully challenge the perpetual peace aspirations of any state leader. However, Kant's remarks on the third preliminary article implies that governments would eventually realize that standing armies also compel rivals to "outdo one another in the number of armed men, which knows no limit."[115] They would thus realize that an unlimited growth of standing military forces could not lead to peace but to a constant and eventually worldwide armistice that would inevitably fall into ever more destructive wars. Kant's analysis thus anticipated the concepts of the security dilemma and security paradox developed more fully in the post–World War II International Security Studies literature.[116] Furthermore, Kant highlights the moral problem of imposing insecurity on individuals in the name of state security—that is, that

> being hired to kill or be killed seems to involve a use of human beings as mere machines and tools in the hands of another (a state), and this cannot well be reconciled with the right of humanity in our own person.[117]

Kant's reference here is to the categorical imperative formulated as "the right of humanity in our own person," and he intends it to function as the practical moral and political grounds for the abolition of standing armies. It seems to follow that any concrete action by states to abolish standing armies, or in more recent terminology a "general and complete disarmament," is an indicator of their likely embrace of common security.

Kant's sixth preliminary article relates another necessary condition of the process of perpetual peace understood as common security. It prescribes that "no state at war with another shall allow itself such acts of hostility as would have to make mutual trust impossible during a future peace."[118] In this article, Kant acknowledges that states at war cannot avoid inflicting mutual damage or harm; however, he insists that some forms of hostility make it practically impossible for state adversaries to conclude peace. The additional necessary condition for shifting from national to common security that Kant identifies in the sixth preliminary article is mutual trust:

> For some trust in the enemy's way of thinking must still remain even in the midst of war, since otherwise no peace could be concluded and the hostilities would turn into a war of extermination. . . . From this it follows that a war of extermination, in which the simultaneous annihilation of both parties and with it all right as well can occur, would let perpetual peace come about only in the vast graveyard of the human race. Hence a war of this kind, and so too the use of means that lead to it, must be absolutely forbidden.[119]

Kant implies that wars of extermination are impossible to avoid unless states realize, as Sisela Bok states, that "their own survival depends on that of all nations."[120] Accordingly, Kant argues that state adversaries can arrive at peaceful relations only if they maintain some measure of mutual trust in their commitment to exercise military restraint. The establishment of mutual trust among enemy states at war requires an acceptance that they share a common interest in peace and, from an ontological security standpoint, at least one core common identity trait that transcends other important identity differences.[121] Indeed, it strongly suggests a mutual acceptance of the need to abandon any kind of ideological fundamentalism inimical to their objective security interests.[122]

Kant's effort in PP to identify the necessary conditions of a process of perpetual peace concludes with a discussion of the relationship between domestic and international political conditions conducive to security. In domestic society, Kant contends that a state's civil constitution is a formal social contract that establishes a "lawful condition" among fellow citizens that abolishes any state of war among them. In international society, sovereign republican states that construct for themselves a (con)federal order[123] have thereby embraced a social contract that establishes for them a lawful condition. As a result, each person and state within this interstate (con)federation are simultaneously "in accord" with the rights of all other citizens (*jus civitas*), the rights of all other states (*jus gentium*), and the rights of all individuals as members of the human race independent of political jurisdiction (*jus cosmopoliticum*).[124] The pacific federation thus institutes a common security arrangement of all member states and peoples and, if universalized, all humanity as well.

One might object that Kant's account of the pacific federation is strictly limited to peoples or states that have been constituted according to republican principles.[125] Thus, the objector might claim that a common security arrangement among republican and non-republican states is not something that Kant could have envisioned, and this in turn would make the invocation of Kant's model problematic for my account. This objection recalls John Rawls's Kantian-inspired *Law of Peoples*, which envisions a realistic utopia, a society of liberal peoples, constituted according to principles of justice as fairness applied internationally.[126] Rawls imagines that this society of liberal peoples could coexist with non-liberal but "decent hierarchical societies" that honor fundamental human rights without providing for a full range of civil liberties for their citizens.[127] However, Rawls does not imagine that a society of liberal peoples could cohabit with or tolerate outlaw states, which disregard fundamental human rights and refuse to comply with international agreements.[128] Rawls might echo and amplify the objection that the possibilities of constructing a common security arrangement with liberal, non-liberal, and illiberal states are beyond Kantian theory's capabilities. The most it can do is ground a collective security interpretation of a liberally oriented security community.

It should be obvious to many readers by now that the common security/security community interpretation of Kant's PP is an alternative to that offered by the democratic peace literature.[129] This is not to say that a federation of democratic states might not count as a security community or exhibit some features of common security relationships; rather, it is that Kant's account should not be misconstrued as an endorsement of a collective security regime of democratic states in which warfare remains for them a morally justified means of confronting rogue state actors or illiberal alliances. If, in fact, liberal democratic states avoid war with each other but fight wars against their ideological enemies,[130] then liberal democratic collective security practices would run afoul of the Kantian moral imperative against war and PP's practical imperative against interfering in the constitutions of other states (i.e., Preliminary Article 5).[131] This is to say, the democratic peace would count as a type of ontological security arrangement among like-minded liberal states in which "security against" remained integral to their politics. But, this is just the kind of security orientation that produced the moral impasse on nuclear defense and deterrence policies that a new twenty-first-century nuclear ethics would seek to overcome.

CONCLUSION

If the previous section has correctly interpreted Kant's account of perpetual peace as prescribing a common security approach for state adversaries, then we are close to articulating a common security imperative that might anchor a new twenty-first-century ethics. This section offers some concluding remarks on the co-constitution of domestic and international security relationships as Kant relates in his definitive articles for perpetual peace.

One necessary element of this co-constitutive dynamic, as PP's first definitive article relates, is concerned with a state's domestic political society: "the civil constitution of every state shall be republican."[132] Ideally, each state in the international order would have a republican constitution. Until that ideal is realized, the nonideal challenge in the pursuit of an indefinite peace is, first of all, the constitution of common security relationships within each existing republican state. This is to say, in a world of republican and non-republican states, Kant's federation cannot be successfully formed unless each republican state's domestic order encodes common security norms among its citizens if they are polarized into hostile ideological camps. Domestic law alone is insufficient to maintain the required climate of "security with" among individual citizens and camps so divided. It also requires an "enlightened" populace whose social interactions are predicated on the free and rational exchange of ideas and policy proposals.[133] We might thus imagine that such a populace would choose to interact around the following maxims of political discourse as related by the Kantian philosopher, Onora O'Neill:

1. To think for oneself, which does not imply a strong individualism but rather the toleration of a plurality of opinion and a refusal to defame or harass another for their views;

2. To think from the standpoint of everyone else, and to see one's initial judgments on issues of import through the eyes of others; and
3. To think consistently, which implies an unending toleration.[134]

In O'Neill's view, the fact of a "plurality of opinion" and the temptation to "defame or harass another for their views" suggests the ineliminable condition that Kant called "unsociable sociability" among a republic's citizens and their respective ideological camps. For Kant, this unsociable sociability is "healthy" insofar as it "compels our species to discover a law of equilibrium" that might resolve conflicts peacefully.[135] This law of equilibrium seems to find expression in the dual role of "thinking for oneself" and "thinking from the standpoint of everyone else." Consequently, O'Neill's three maxims emphasize the importance of individual autonomy in domestic political contexts where "thinking for oneself" is a rational and empathetic process of weighing differences of viewpoints fairly and justly. An enlightened citizen has escaped intellectual "minority" if they exercise their own reasoning on the merits of any proposed policy aiming at the public good.[136] Nonetheless, that exercise of reason includes the respectful consideration of others' viewpoints and experiences. Second, enlightened citizens are capable of maintaining differences of opinion without regarding them as ideologically rooted threats to the common good. Their empathetic exercise of reason suggests an acceptance of "security with" and a corresponding rejection of ideological fundamentalism within domestic political interactions. This seems to be the essence of O'Neill's use of the term "toleration," and it becomes a groundwork for applying this culture of common security into the international realm.

Accordingly, a republican state constituted by a culture of common security is positioned to advance its right, according to PP's second definitive article, to construct "a federalism of free states."[137] Kant sees this federal relation as approximating a civil constitution among these states but without establishing a "state of nations" that would deprive each of its sovereignty.[138] This is to say, this federal relation enables each state as its own political community to remain autonomous and also part of what Ion calls a "cosmopolitan community."[139] As a result, the relational dynamic of the pacific federation mirrors to a great degree that of autonomous and enlightened citizens, and the communities to which they belong, within a republican state. For states as well as citizens, the right of war would be "unintelligible" as it would undo the lawful condition and subvert their security.[140]

Moreover, as PP's third definitive article suggests, the duty of each republican state in this federation to honor the cosmopolitan right of hospitality ensures against the effects of any residual ideological fundamentalism and, consequently, facilitates and sustains the culture of common security across state borders. By honoring the rights of any foreign guest to seek opportunities for economic and cultural exchange, each republican state affirms each individual's "right of possession in common of the earth's surface on which, as a sphere, they cannot disperse infinitely but must finally put up with being near one another."[141] Kant's idea concerning each individual's "right of possession in common of the earth's surface" anchors the right of foreign travel and trade, given the unequal distribution of natural resources across our world. Thus, the ground of the "right of possession in common of the earth's surface" combines with the right to federalism of free states and the corresponding duties to avoid war and never engage in wars of extermination leading to the inference that the security conception that can best anchor an international security ethics, and more specifically a twenty-first-century nuclear ethics, is that of common security in the tradition of Deutsch and Palme.

In closing, none of the foregoing establishes anything other than the relevance and salience of this anchoring concept of common security for a twenty-first-century nuclear ethics. The articulation of such a nuclear ethics consistent with this concept awaits the efforts of chapter 6. Between that chapter and this one, two other chapters explore how nuclear-armed liberal democracies encounter problems of ontological and moral incoherence (chapter 4) and why the moral imperative to abolish nuclear weapons must find a corresponding morally responsible process (chapter 5). By these efforts, the book will provide further reason for embracing the alternative concept of common security and its corresponding maxims of nuclear ethics.

NOTES

1. See, for example, Hehir (1986); National Conference of Catholic Bishops (1983); Nye (1986); Payne and Payne (1987); Quinlan (1984 (2009)); Ramsey (1968 (2002)); Walzer (2015 (1977)).

2. See, for example, Bok (1988); Dummett (1986), Goodin (1985); Lee (1985); Schell (1984).

3. For recent accounts on the moral dilemmas of nuclear defense and deterrence, see, for example, Doyle II (2015) and Heinze (2016).

4. One of the best accounts of this is Craig and Ruzicka (2013). See also Walker (2012, 184–91).

5. See, for example, Nolan (1999); Sauer (2005); Wittner (2009).

6. Deutsch et al. (1957).

7. Palme (1982b).

8. For an early post–Cold War era treatment, see Cronin (1999).

9. Krauthammer (1990/1991).

10. I thank Brad Roberts for suggesting the term "polycentricity."

11. "Nuclear Posture Review: February 2018" (Washington, D.C., 2018), 6.

12. The literature on the evolution of U.S. nuclear defense and deterrence policy across the Cold War and post–Cold War eras is too large to cite in its entirety. For

works which focus significantly or mostly on Cold War nuclear policy, see for example, Ambrose and Brinkley (1997); Gavin (2012); Lee (1993); Smoke (1993); Tannenwald (2007); and Walker (2012). For works which focus significantly or mostly on the post–Cold War period of nuclear policy, including the threat of nuclear proliferation, see, for example, Bernstein (2014); Brands (2016); Campbell, Einhorn, and Reiss (2004); Larsen and Kartchner (2014); Morgan (2003); Morgan and Paul (2009); Roberts (2016); Solingen (2007); and also Walker (2012).

13. Walker (2012, 109–12).

14. See Norris and Kristensen (2010).

15. Information in Table 3.1 and the related discussion is sourced from Morgan and Paul (2009); Roberts (2016); Walker (2012).

16. Tom Sauer disagreed that the Bush 41 administration "embraced the hope of an enduring peace with Russia" on the grounds that it did not take Russia into the European security architecture on an equal basis nor did it want to get rid of NATO. Private correspondence, September 22, 2018.

17. Morgan and Paul (2009, 265).

18. On Saddam Hussein's use of chemical weapons during this period, see BBC News (2003).

19. One concern over the use of nuclear deterrence to counter chemical or biological weapons attacks is found in Sagan (2000).

20. The White House (2002); Roberts (2016, 21).

21. Roberts (2016, 110–17).

22. Obama (2009); Thakur (2015, 5).

23. U.S. Department of Defense (2010); Roberts (2016, 31–32).

24. Roberts (2016, 35–36).

25. 2010 Review Conference of the Parties to the Treaty on the Non-Proliferation of Nuclear Weapons (2010); Huntley (2010); Doyle II (2015, 74–77).

26. See, for example, Morgan and Paul (2009); Roberts (2016, chapters 4 and 5); Wang (2012).

27. Roberts (2016, chapters 2 and 3). On weak state deterrence, see Arreguin-Toft (2009).

28. Snyder (2016).

29. Gladstone (2016); Sanger and Gladstone (2017).

30. Landler (2018).

31. Acheson (2015a, b); Baker (2010).

32. Roberts (2016). My thanks to Brad Roberts for his feedback on earlier drafts of this chapter.

33. Roberts biography is related on Stanford University's Center for International Security and Cooperation's website: https://cisac.fsi.stanford.edu/people/brad_roberts Accessed 9/5/2018.

34. Roberts (2016, 30).

35. Roberts (2016, 52).

36. Roberts (2016, 58–69).

37. Roberts (2016, 75–82).

38. Roberts (2016, 83–84).

39. Booth and Wheeler (2008, 141).

40. Donaldson (1985); Lee (1985).

41. Booth and Wheeler (2008, 173–77).

42. Roberts (2016, 95).

43. Roberts (2016, 95–96).

44. Baker and Choe (2017); The Century Foundation (2018).

45. Hass (2018).

46. Office of the Secretary of Defense (2018, I).

47. Office of the Secretary of Defense (2018, 6).

48. Office of the Secretary of Defense (2018, 8–10).

49. Office of the Secretary of Defense (2018, 11, 31).

50. Office of the Secretary of Defense (2018, 11–14).

51. Office of the Secretary of Defense (2018, vii–viii).

52. Rose (2018).

53. Payne (2018).

54. De Luce and Gramer (2017).

55. I borrow this term from Wade Huntley, who in private correspondence correctly raised this larger point. My thanks to Wade for his insight.

56. Arms Control Association (2018).

57. Kuhn (2018, 249).

58. Davis (2018); Erlanger and Bennhold (2019); Herszenhorn and Bayer (2018).

59. Kuhn (2018, 249–50).

60. Booth and Wheeler (2008, 298).

61. Buzan (1987).

62. Palme (1982b). Also quoted in Booth and Wheeler (2008, 139).

63. Buzan (1987, 102). For more recent treatments of cooperative security, see Davidson (2010); Harding (1994); Molder (2011); Zellner (2010).

64. Deutsch et al. (1957). For later treatments of security communities, see Adler and Barnett (1998); Bellamy (2004).

65. Deutsch et al. (1957, 5).

66. Deutsch et al. (1957, 30). The seven instances of failure of amalgamated security communities that Deutsch identified are Metropolitan France from 1789 to 1871; the United States and Confederate States of America in 1861; Norway and Sweden in 1905; England and Ireland in 1918; Austria-Hungary in 1918; Spain, Catalan, and Basque in the 1930s; and France and Algeria in the 1950s.

67. Deutsch et al. (1957, 65–67).

68. Deutsch et al. (1957, 60).

69. Fisher and Bennhold (2018).

70. Davidson (2010, 20).

71. Zellner (2010, 64).

72. Zellner (2010, 65).

73. Molder (2011).

74. *The Economist* (2018).

75. Palme (1982a, 1982b).

76. Holst (2011).

77. Palme (1982a, 6).

78. Schelling (1966, 1–34). This vulnerability of peoples in the context of state incapacities factors indirectly into Thomas Schelling's discussion of the diplomacy of violence.

79. Palme (1982a, 7).

80. Palme (1982a, 13).

81. Palme (1982a, 7).

82. Palme (1982a, 8–9).

83. Palme (1982a, 9).

84. For an extended discussion of this topic in relation to conventional arms, see Bower (2017, especially chapters 4 and 5).

85. Palme (1982a, 10).

86. Palme (1982a, 11).

87. Palme (1982a, 11).

88. For a critique of Palme's contribution to the common security framework, see Buzan (1987, 267–68).

89. Buzan and Hansen (2009, 136).

90. Buzan and Hansen (2009, 137).

91. Booth and Wheeler (2008, 159).

92. Booth and Wheeler (2008, 163, 209–12).

93. The notion of ideological fundamentalism is borrowed from Wheeler (2018, 93–96).

94. See, for example, Mitzen (2006); Steele (2008).

95. Steele (2008, 2–3).

96. Steele (2008, 13).

97. Table 3.2 is a modification of the table related in Steele (2008, 52).

98. Steele (2008, 94–113).

99. Steele (2008, 102–6).

100. Quoted on Steele (2008, 95).

101. Steele (2008, 114–47).

102. For an in-depth comparative case study of the relationship between state-identity conception and nuclear proliferation, see Steele (2008, 114–47).

103. Identify formation seems to require acts of differentiation from otherness in some mode. However, that otherness might arise from various sources: external actors or, in some cases, a stigmatized element of one's past identity. I thank Laura Considine for bringing this issue to my awareness.

104. Booth and Wheeler (2008, 65).

105. Wheeler (2018, chapter 3).

106. Booth and Wheeler (2008, 65–70).

107. Campbell (1992, chapters 6 and 7).

108. Campbell (1992, 136–40).

109. Booth and Wheeler (2008, 114–18).

110. Booth and Wheeler (2008, 118–23).

111. Kant (1795, 1996d, 491).

112. Ion (2012). A compatible security conception is Karin Fierke's "security clusters," which is developed in Fierke (2015, 45–48). For a contrasting view on Kant's theory of international politics and peace, see Nardin (2017).

113. Kant (1795, 1996d, 317–18).
114. Kant (1795, 1996d, 318).
115. Kant (1795, 1996d, 318).
116. Booth and Wheeler (2008, chapters 1 and 2).
117. Kant (1795, 1996d, 318).
118. Kant (1795, 1996d, 320).
119. Kant (1795, 1996d, 320).
120. Bok (1988, 23).
121. On the subject of trust among states, see Booth and Wheeler (2008, chapter 9); Kydd (2005); Larson (1997); Rathbun (2012); Wheeler (2018).
122. Booth and Wheeler (2008, 165).
123. Kant (1795, 1996d, 325–28).
124. Kant (1795, 1996d, 322).
125. I thank Wade Huntley for raising this issue with me in private correspondence (December 21, 2018).
126. Rawls (1999).
127. Rawls (1999, 62–71).
128. Rawls (1999, 80–81).
129. Ion (2012, 3, 5–39); Buzan and Hansen (2009, 167–68). See also, for example, Doyle (1983, 1986); Lake (1992); Maoz and Russett (1993); Russett et al. (1993).
130. Doyle (1983).
131. Kant (1795, 1996d, 319).
132. Kant (1795, 1996d, 322).
133. Kant ((1784) 1996a).
134. O'Neill (1986, 540–47). See also Ion (2012, 23–24).
135. Ion (2012, 26).
136. Kant ((1784) 1996a).
137. Kant (1795, 1996d, 325); Ion (2012, 45).
138. Kant (1795, 1996d, 326).
139. Ion (2012, chapter 3).
140. Kant (1795, 1996d, 328).
141. Kant (1795, 1996d, 329).

Chapter 4

The Ontological and Moral Incoherence of Liberal Nuclearism[1]

This book's second chapter explained how the Cold War era stalemate among nuclear ethicists on the moral justifiability of nuclear defense and deterrence was based on their fundamental disagreement over which set of moral values or imperatives were ultimately decisive. Specifically, it argued that the moral justifications of nuclear defense and deterrence privileged considerations of national survival and security in the international order over considerations of justice among states and peoples. Furthermore, it argued that moral condemnations of these policies privileged considerations of the justice of human security and noncombatant immunity over widely held and historically rooted rights of national self-defense. The third chapter related how this stalemate continued into the post–Cold War era, which most recently has been defined by the Doomsday Clock's setting at "two minutes to midnight." Finally, it argued that the condition of "two minutes to midnight," which represents global existential insecurity, makes it morally imperative for nuclear adversaries to restructure their relations around a common security conception. Such a restructuring would provide greater opportunity for a just international order necessary to construct a durable peace.

This fourth chapter examines more deeply the political and moral consequences of liberal nuclearism, which is constituted by a liberal democratic state's commitment to nuclear defense and deterrence policies as grounded on security conceptions that, in the terminology introduced in chapter 3, favor "security against" dynamics. It addresses the following questions: Have nuclear-armed liberal democratic states suffered significant and perhaps irreversible damage to their domestic institutions and security policy arrangements as a result of their nuclear defense and deterrence postures, especially since the end of the Cold War? If so, does that damage constitute a crisis in liberal ontological security such that (1) any claim by a nuclear-armed democracy to being "liberal" is no longer coherent and, for that reason, (2) any claim of moral authority or legitimacy in security policy by a nuclear-armed democracy is also incoherent?

This chapter explores these questions by using John Rawls's 1999 monograph, *The Law of Peoples* (LP), as a central organizer for analysis. LP counts as one of the chief exemplars of a liberal internationalism that comports with one central national identity conception embraced by the United States and its closest democratic allies.[2] This chapter argues that the U.S. commitment to nuclear defense and deterrence has damaged and weakened its liberal institutions as described by Rawls, constituting a crisis of incoherence for its liberal ontological security and its moral authority in security policy. If liberal nuclearism has induced or will induce similar crises in other nuclear democracies—for example, the United Kingdom, France, India, and Israel—then it will become clearer that the best approach to securing liberalism as a valued set of political and social arrangements is through nuclear disarmament.

The chapter proceeds as follows. The first main section sets the stage for the chapter's main analysis by briefly reviewing how the concept of ontological security shifts the referent of security away from a narrow understanding of protecting territorial integrity and regime stability to national identity and its corresponding social and political arrangements. A conception of ontological coherence is inferred as that condition in which a national identity conception holds together if a state's political practices are consistent with its foremost identity narratives. By contrast, ontological incoherence is the product of profound inconsistencies between identity narratives and political practices. Such incoherence is a grave threat to a state's ontological security and the moral authority on which it rests. The second section interprets Rawlsian liberalism as a core narrative of liberal democratic identity in the post–World War II period. The third section argues that Rawls's defense of nuclear defense and deterrence—following Michael Walzer's defense in *Just and Unjust Wars*—constitutes a rupture in liberal moral thought.[3] The rupture emerges as a contradiction between the liberal insistence on the rule of law over the rule of force and the requirement of nuclear defense and deterrence policy that the rule of force must be decisive when liberal states confront so-called "outlaw states."[4] The fourth section draws inferences on how a commitment to national security by nuclear defense and deterrence transforms liberal peoples into outlaw states. This transformation is a sufficient condition of incoherence for liberal ontological security and the moral narratives on which the former rests. The chapter concludes with brief remarks on the Trump administration as the latest manifestation of incoherence of American national identity and its liberal moral narratives.

STAGE SETTING: LIBERAL ONTOLOGICAL
(IN)SECURITY AND MORAL (IN)COHERENCE

The literature on ontological security in International Relations generally contends that state actors are concerned with the security of their perceived "being"—that is, their national identity as expressed by an honored set of historical, cultural, and political narratives—more than their physical survival *per se*.[5] Jennifer Mitzen puts it thusly: "Ontological security refers to the need to experience oneself as a whole, continuous person in time—as being rather than constantly changing—in order to realize a sense of agency."[6] Individuals seek security in "who they are," so they can marshal the motivations required for socially significant instances of choice and action. And part of this process is the finding of one's social place in a national collective, which can enhance one's sense of stability and predictability in future interactions. Finding one's own place also provides a sense of individual purpose and meaning as a valued member of a national society. Individual agency exercised for the collective (national) good thus enables collective agency to achieve that good, which in turn solidifies and validates the honored national identity conception affirmed by a state's citizens. Hence, the inability to realize collective agency can thus constitute a kind of existential incapacity or "death" that must be avoided in order to maintain a satisfactory level of security.

The recognition of national identity as a central referent of security marks a significant departure from mainstream IR realist accounts. For IR realists, the drivers of state security policy are almost always independent of national identity conceptions and their function in domestic and international politics. Rather, for IR realists, states' security policy is centrally focused on addressing unfavorable imbalances of power in regional or global political environments, which, if ignored, can lead to aggression, military defeat, and potential death of the state.[7] Contributors to the ontological security literature acknowledge that state actors facing immediate threats of aggression will generally put physical survival above the preservation of other values. However, they highlight more than a few instances of supposed "irrational" behavior of state actors who risk survival in the pursuit of interests related to national identity.[8]

How are such anomalous cases explained? According to Brent Steele, the willingness of state actors to embrace significant risks to their survival is, as hinted at earlier, often linked to the psychological need to avoid or resolve experiences of national dishonor or shame.[9] National dishonor or shame counts as a condition of ontological incoherence, and it arises when one or more foreign policies (or their outcomes) are taken as inconsistent with cherished national identity narratives. The inconsistency attributed to these foreign policies in relation to national identity narratives constitutes a rupture or breach of the national "self." A rupture of the national "self" constitutes a national identity crisis of the first order, which exposes a critical social psychological component to national security excluded from IR realist accounts. In such conditions, it is not infrequent that leaders or members of the national community object that such actions "do not reflect who we are." If the questionable policy is still a matter of legislative or administrative debate, then the objection of "this is not who we are" is prospective and a way to avoid shame and an existential crisis of identity. Thus, the Melians in Thucydides's *Peloponnesian War* deliberated on the Athenian ultimatum to surrender and chose to endure utter defeat to avoid the greater shame of betraying their honor.[10] Another case is that of Belgium, which was discussed extensively in the previous chapter.[11]

So far, the chapter's discussion has assumed that states have one identity conception of supreme value that is the focal object of security. However, ontological insecurity and incoherence can also be produced by diverse efforts to preserve competing and sometimes contradictory national identities. According to Amir Lupovici, *ontological dissonance* is a kind of ontological incoherence that occurs when states hold multiple identity conceptions and the means of addressing threats to each of their identities cannot escape a mutually destructive dynamic.[12] In such conditions, Lupovici argues that the management of state identity conception A might undermine a contrasting management strategy for state identity conception B, and vice versa. States must then choose between different cherished identities, which constitutes a moment fraught with varying degrees of instability and insecurity. For Lupovici, one way of addressing ontological dissonance is to discursively shift from the identity conception that appears more offensive toward the more acceptable one. Another is to change policy and thus change state behavior consistent with the unoffensive identity conception. Last, a state may choose to avoid the dissonance by using mechanisms to screen out exposure to the information that sustains the identity threat of greatest concern. According to Lupovici, avoidance permits state actors experiencing ontological dissonance to revalidate their identity rather than change their public discourse or behavior. Through avoidance, state actors attempt to mitigate the threats to one of their identity conceptions and endure some measure of ontological incoherence without it reaching crisis levels.[13]

Lupovici's case study of the Israeli-Palestinian crisis examines Israel's struggle with ontological dissonance during the Second Intifada (2000–2005) when Palestinian terrorism simultaneously threatened Israel's Jewish, democratic, and security-provider identities.[14] Israel's policy to protect its Jewish identity was to build a security barrier to exclude the Palestinians. However, that policy undermined any strong claim Israel had to *being* an inclusive liberal democracy. Ultimately, the inability of the security barrier to put a complete stop to Palestinian terrorism—then and now—threatens Israel's security-provider identity, which exacerbates the ontological security dilemmas over its Jewish and liberal identity conceptions. Lupovici argues that Israel's decision to emphasize the security barrier reveals a choice to avoid the threatening information concerning the democratic and security-provider identities in favor of the Jewish state identity.

Lupovici's concept of ontological dissonance finds a specific application relative to the great powers in what Bruce Cronin calls the paradox of hegemony.[15] Examining the United States and its relationship with the United Nations, Cronin finds that Washington is pulled between two fundamentally different national identities—global great power and hegemonic world leader. As a global great power, Cronin argues that the United States must pursue its national interest in ways that comport with IR realist analysis. However, Cronin does not define "hegemonic actor" as do the realists in terms of regional or global domination.[16] For him, a hegemonic actor acts to stabilize and secure the international order through the global institutions that it has helped to construct. To do so, the hegemonic actor must restrain its pursuit of self-interest and work to sustain the legitimacy of institutional relations that might reflect inequalities of power but that do not reflect any state's subordination to the hegemon. Thus, there is a tension for a global great power such as the United States "between parochial interest and international responsibility . . . [which] can be called the 'paradox of hegemony.'"[17] If the global great power fails "to act within the boundaries established by its role [as world leader within international institutions], the credibility of the institutions and rules it helped to establish weakens. . . . When these [international] organizations are undermined, the legitimacy of the international order is threatened. If this persists over time, the hegemonic order declines."[18]

Cronin's analysis of the means by which post–Cold War–era U.S. administrations addressed this paradox strongly suggests a conclusion that is consistent with this book's third chapter: namely, that the United States privileged its identity as world leader until it faced new threats from state actors and growing unconventional threats from non-state actors. Focusing solely on the Bush 41 and Clinton administrations, Cronin argues that the former excelled in reaffirming the U.S. role as world leader in 1990–1991 by organizing a strong coalition of states through the United Nations against Saddam Hussein's Iraq and its aggression into Kuwait. During the Clinton era, the United States tended to emphasize its national interest at the expense of its role as a world leader. For instance, it acted unilaterally to manage a defiant Iraq through airstrikes without the consultation or approval of other United Nations Security Council members. Such actions weakened the U.S. position within the Security Council, and it arguably set the tone for greater unilateralism by the United States during subsequent administrations. Thus, just as Lupovici's account highlighted Israel's choice to privilege its Jewish identity, Cronin's analysis suggests that the United States, insofar as it pursues its global great power identity over its world leader identity, will find that the latter will atrophy. This is because "a systemic leader cannot act as a representative of the broader community if it acts in defiance of the rules of the community."[19]

In short, the ontological security approach can be understood as a set of claims concerning the ontological (in)coherence of a state if its behavior is contrasted with a sole privileged national identity conception (Steele) or in light of competing national identities, each of which suffer some kind of critical threat (Lupovici; Cronin). If this approach is applied to the questions of nuclear defense and deterrence raised in chapters 2 and 3, it can explain why some states acquired nuclear weapons after the momentous decisions by the United States and former Soviet Union to nuclearize. One notable case is that of France, which had suffered humiliation twice in two decades: that is, by its defeat at the hands of and surrender to Nazi Germany in 1940 and after World War II by its marginalization in world status as the United States and Soviet Union arose as the world's superpowers. Once the Soviets and the British had tested nuclear devices, the French believed it imperative to do anything necessary to reclaim their "grandeur." And, for French leaders, this required nuclear weapons.[20] In 1955, then Premier Edgar Faure was quoted as saying:

> I ask myself whether France should forego the right she has to be always in the first rank. . . . I think I shall reject the negative solution, that of giving up [the pursuit of nuclear weapons]. France cannot remain in the category of "great powers in reduced circumstances."[21]

A few years later, in light of the perceived need to nuclearize to hedge against the prospects of West German rearmament as part of the Allied containment strategy against the Soviets, Beatrice Heuser quotes an unnamed French official who had claimed that

> the metaphysical survival of France . . . is the most important; the moral, political, historical annihilation would be seen as worse than only the physical destruction. France must be prepared to risk the latter to save her honour, save her identity.[22]

Taken together, these two quotations are illustrative of Steele's account of shame avoidance and they suggest an ontological security imperative that seemed to override competing prudential imperatives to keep France nonnuclear and thereby not provide the Soviets with a new nuclear adversary. For French leaders, the identity of "great power," which after the Soviet nuclear test of 1949 was newly defined by the possession of nuclear weapons, could not be lost or abandoned without the humiliating loss of what it meant to be "France."

The aforementioned cases identify another distinctive feature of ontological security-seeking: the role of ethics or morality in the construction and securing of a privileged national identity conception. It has been suggested that state actors are driven (as are individuals) to construct coherent self-conceptions and then use biographical narratives and well-planned action or routines in their maintenance.[23] According to Steele, the content of those narratives is value-laden.[24] This is to say, identity narratives emphasize the character virtues that a national people claim to possess, the noble objectives that they claim to seek, and the rights and duties they claim to bear in the pursuit of these objectives. Accordingly, national honor is taken as a cardinal moral virtue linked intimately to these objectives, rights, and duties, and therefore it must be preserved against any compromising efforts. Likewise, instances of national shame or humiliation constitute a grave dishonor and a moral defeat for state actors, insofar as they mark a rupture in the coherence of the people's identity conception. If this analysis is correct, then it explains how the French appeal to its "metaphysical survival" is a rational and retroactive support for its decision to nuclearize, the Belgian defense of its identity as a historic neutral power in light of its "rape," and Melos's futile resistance in the name of justice in the face of Athens's overwhelming power.

The foregoing has argued that a state risks ontological and moral coherence if it acts to buttress one of its cherished identities at the expense of its other ones. The next question is if this risk-taking strategy is rational. While IR realist accounts would deny that such a strategy is rational, Steele argues that it can be. For instance, NATO's 1999 operation in Kosovo, which aimed at preventing the ongoing genocide against ethnic Albanians by the Serbs, can be taken as rational if understood as an attempt by its individual alliance parties to address diverse national identity crises produced by the shame of past failed policy. According to Steele, the United Kingdom used its participation against Serbia and its leader Slobodan Milosevich, whom London had associated with Hitler, to address its memories of guilt over its pre–World War II appeasement with the Nazis at Munich in 1938. Similarly, Germany used its participation to confront its Nazi past and thereby reenter the community of legitimate powers. The United States appeared to have different demons it needed to confront—specifically, its shame at Washington's inaction during the Rwandan genocide.[25] In the latter case, Cronin's account can explain how Clinton's inaction on Rwanda intensified U.S. ontological dissonance in terms of "great power," "world leader," and "liberal democracy."

The case of NATO member states undertaking humanitarian rescue operations in Kosovo for the purpose of ontological security-seeking raises new questions about the adequacy of mainstream IR realist explanations of collective security operations. Certainly, a liberal NATO alliance is interested in preventing the resurgence of autocratic or otherwise Nazi-like groups willing to commit grave atrocities to further their nationalist ambitions. Even so, Steele's and Cronin's accounts, if correct, suggest strongly that NATO's interest in preventing further genocide was sublimated to its member states' distinct ontological security interests. This inference in turn reinforces a claim from chapter 3 that ontological security-seeking could be a potential source of opposition to the adoption of a common security approach to resolving nuclear security dilemmas. This claim highlighted a significant challenge to common security approaches, which chapter 3 argued were required to transform global or regional great power adversarial relations away from "security against" and toward "security with" postures. Through careful and patient diplomatic effort, nuclear-armed adversaries are challenged to acknowledge and prioritize the common interest in survival and security, which the possession and threatened use of nuclear weapons puts at constant existential risk. The recognition of this common security interest is mutually constitutive with the mutual effort at empathy and trust-building within an institutional framework. If these practices are effective, then it eventually leads to a mutual redefinition of each other's national identities, which might be generally expressed as shifting from "they-as-enemy" to "they-as-partner."

And yet, chapter 3 also states that common security is a liberal conception, one that appears to have at least some roots in Kant's conception of perpetual peace.[26] This chapter's original questions are now raised, but they might profit from some reframing: (1a) if the ontological security of a liberal democratic state requires nuclear deterrence, does not that same reliance produce a rupture in their own identity conceptions? (2a) If reliance on nuclear deterrence by a liberal democratic state produces a rupture in their identity conception, then can any moral justification of liberal nuclearism in the name of its national security remain coherent on moral grounds? The following section begins the critical exploration of these questions by reviewing the political liberalism of John Rawls as an exemplar discourse of liberal American identity.

RAWLSIAN LIBERALISM: AN EXEMPLAR OF CONTEMPORARY LIBERAL DEMOCRATIC IDENTITY

John Rawls is widely recognized as the latter twentieth-century philosopher most responsible for resurrecting political philosophy after its marginalization by the rise of analytic philosophy in the early 1900s.[27] Rawls's main works, *A Theory of Justice* (TJ) in 1971 and *Political Liberalism* (PL) in 1996 arguably advance together one of the most influential accounts of the features of domestic liberal democratic political order.[28] These core features are *justice as fairness, the equal liberty of citizens, public reason*, and *constitutionalism*. For Rawls, each of these concepts are necessary and jointly sufficient for a domestic society to count as liberal. The concept of justice as fairness describes two basic principles of liberal society: that each person shall have an equal right to the most extensive scheme of basic liberties compatible with a similar scheme for others, and that social and economic inequalities should be arranged to the greatest benefit of the least advantaged and attached to offices and positions open to all under conditions of fair equality of opportunity.[29] These two principles of justice place the equal liberty of citizens at the heart of liberal society. From there, Rawls claims that public reason is the exercise of reason by the representatives of free and equal citizens over matters of constitutionality and basic justice. It is the exercise of public reason that makes a constitutional democracy a deliberative democracy.[30] These representatives include popularly elected legislators and executive officers as well as candidates for such offices and the judiciary.[31] It is through these constitutional devices that liberal democratic society implements (however imperfectly) the ideal of procedural justice that satisfies all the basic requirements of the principle of equal liberty of citizens.[32] These devices include a bicameral legislature, separation of powers mixed with checks and balances, and a bill of rights with judicial review.[33] The degree to which the ideal of procedural justice is operationalized is the degree to which liberal society domestically can validly claim to have reconciled liberal ideals with their social and political realities. And such claims would then become part of the national narratives expressing one of the main, if not the central, identity conception of a liberal democratic state, and upon which its ontological security-seeking is based.

To say that Rawls's account of political liberalism has been widely influential is not to dismiss other competing accounts of liberalism. Judith Sklar, for instance, contrasts the Jeffersonian liberalism of rights, the Emersonian liberalism of self-development, and what she calls the "liberalism of fear" that appears increasingly to characterize contemporary liberal society.[34] For Sklar, each of these liberal accounts concur that individual persons ought to be free to pursue their interests in a way that is compatible with the freedom of all other individuals. Thus, one of the basic elements of liberalism is the mutually held right of freedom from interference in one's pursuits. However, for Sklar, the contemporary liberalism of fear is distinct from other liberalisms insofar as its core concern is freedom from "the abuse of power and [the] intimidation of the defenseless,"[35] which arises when domestic or international political society suffers from the cruelty produced by stark power asymmetries. Whereas the Jeffersonian and Emersonian liberalisms rest on some conception of the *summum bonum*, or the greatest good that individuals and groups can achieve if they were unfettered by government constraints,[36] the liberalism of fear rests on a concrete conception of the *summum malum*, or the greatest evil that individuals and groups might suffer at the hands of the powerful. Accordingly, Sklar defines "cruelty" as the deliberate infliction of physical and emotional pain upon the weak by the strong in order to achieve some political end.[37] And it is the "systematic fear" of experiencing this cruelty at the hands of stronger actors, which "makes freedom impossible."[38] As I see it, Rawls's concern for justice as fairness and constitutionalism, while not advanced in terms of the *summum malum*, is directly related to the prevention of abuses of power and any stronger party's exercise of cruelty for their own advantage. This will become clearer later when the discussion focuses on Rawls's concern for "outlaw states."

Rawls's LP applied these features of domestic liberal orders to a prospective international order in which the anarchy condition is tamed by the corresponding features of the democratic peace based on the Kantian concept of the pacific federation among constitutional republics.[39] For Rawls, this Kantian concept is based on "the social contract idea of liberal political conception of a constitutionally democratic regime."[40] Rawls' use of "social contract" assumes an original anarchy condition among diverse national political communities that, despite any existing liberal societies' well-orderedness, produces mutual and existential insecurities for all. In this vein, Rawls recalls that TJ originally imagined the extension of justice as fairness to international law for the limited purpose of judging the aims and limits of just war.[41] The consideration of just war theory as a topic for justice as fairness confirms the assumption of the anarchy condition among adversarial states (or peoples), which arises in the invocation of "social contract."

Indeed, Rawls's worry about the existential insecurity of liberal societies in an anarchical international order is LP's central motivation. Rawls claims that a law of peoples is essential because of the urgent desire to prevent the "great evils of human history" from recurring: unjust war and oppression, religious persecution and the denial of liberty of conscience, starvation and poverty, and genocide and mass murder.[42] In Rawls' view, preventing the recurrence of these great evil requires the constitution of many constitutional republics that, when their institutions are well-established, will cultivate by social habit the respect for peace, human rights, and justice as defined by political liberalism. The second step of preventing great evil is by instituting such habits among liberal republics, thereby instituting what Michael Doyle and others have called the "democratic peace."[43] Indeed, as Rawls states,

> The crucial fact of peace among democracies rests on the internal structure of democratic societies, which are not tempted to go to war except in self-defense or in grave cases of intervention in unjust societies to protect human rights. Since constitutional democratic societies are safe from each other, peace reigns among them.[44]

In short, a law of peoples must take all necessary steps to ensure that the character and structure of liberal order is constructed and preserved against unjust illiberal orders. One part of that effort involves inculcating the habits of democracy, which in the liberal narrative involves restraints on violence except in self-defense or the defense of human rights. The liberal narrative is thus organized around the values of justice, constitutionality, human rights, and liberty. The spread of democracy regionally and then globally involves the promulgation of this narrative and its embrace by other nation-states.

Now, recalling that the composition of the international order remains anarchic, even if a discrete society of liberal peoples can construct a pacific federation, Rawls advances a third step for the prevention of great evil:

> Yet so long as there are outlaw states, as we suppose, some nuclear weapons need to be retained to keep those states at bay and to make sure they do not obtain and use those weapons against liberal or decent peoples. How best to do this belongs to expert knowledge, which philosophy does not possess.[45]

Rawls' third step marks the analytical point of concern for this chapter. Before this critical analysis is undertaken, it is important to note its location in the Rawlsian narrative of the Society of Peoples. To do this, it is important to detail Rawls' accounts of the outlaw state, the liberal commitment to international law and principles, and the corresponding necessity of liberal societies to resort to nuclear deterrence against outlaw states. The first two of these elements have been embraced as part of the liberal identity narrative, and the latter is the foremost security policy against hostile illiberal aggression.

RAWLS'S LP AND THE LIBERAL COMMITMENT TO INTERNATIONAL LAW AND PRINCIPLES

The previous section related that, for Rawls, one feature of a democratic people (or, for my purposes, a democratic state) in relation to a Society of Peoples is its commitment to international law and institutions constructed along liberal principles. Indeed, this commitment is an essential feature of how liberal peoples distinguish themselves from unjust illiberal states. It also counts as a necessary for the establishment of the kind of security community (chapter 3) that corresponds with the democratic peace.[46]

For Rawls, this account of liberal international institutionalism must begin with the Kantian state of nature assumption, that is, all sovereign political societies are at war merely by being near one another.[47] This condition of war is their central motivation to reconstitute themselves domestically as constitutional republics (or liberal peoples) and subsequently enter into a pacific federation with other constitutional republics consistent with the Kantian notion of social contract.[48] Insofar as the Law of Peoples is an application of justice as fairness to the international level, then a commitment to justice as fairness entails a commitment to international law and its principles.[49] For Rawls, the principles of the Law of Peoples specify the following:

1. Peoples are free and independent, and their freedom and independence are to be respected by other peoples.
2. Peoples are to observe treaties and undertakings.
3. Peoples are equal and are parties to the agreements that bind them.
4. Peoples are to observe a duty of nonintervention.
5. Peoples have the right of self-defense but no right to instigate war for reasons other than self-defense.
6. Peoples are to honor human rights.
7. Peoples are to observe certain specified restrictions in the conduct of war.
8. Peoples have a duty to assist other peoples living under unfavorable conditions that prevent their having a just or decent political and social regime.[50]

Two points concerning this list of principles deserve mention. One, these principles are largely restatements of existing international norms codified in current international law, some of which (e.g., honoring human rights) were introduced by liberal internationalists between the nineteenth and twentieth centuries.[51] Second, Rawls admits that these principles are incomplete and that many require further explanation and interpretation.[52] However, Rawls's overall account in LP suggests that if international society were composed solely of well-ordered liberal peoples under a law of peoples, then the fourth, sixth, and seventh principles would be superfluous. The superfluity of these principles would count as a positive affirmation of the universalization of justice as fairness in international relations, and accordingly it would align with the core contention of this book: that the fundamental considerations of justice ought to be the ordering principle of a society of states such that the survival and security of states and peoples is fully realized. Until that time, the full set of principles is necessary to address conflicts that will arise within a mixed international order of liberal and nonliberal societies. In such a mixed international order, liberal societies might become stuck in a series of ethical foreign policy dilemmas across a wide range of security issues unless some normative principle for adjudicating conflicts among rights or duties is specified.[53] Rawls' remark on the fourth principle—that of nonintervention—provides some of that specification insofar as it "will obviously have to be qualified in the general case of outlaw states and grave violations of human rights."[54] This is to say, for Rawls any society or state's failure to uphold this absolute duty forfeits the right of noninterference and warrants the practice of liberal humanitarian intervention.[55]

In this vein, we recall that Rawls' original interest in international law was restricted to assessing the aims and limits of just war, which specifies that the general right to war revealed by this tradition limits the use of armed force to national self-defense.[56] This is expressed above in the fifth principle of the Law of Peoples. Rawls recognizes that other political societies in the world have the right of armed self-defense, but we have seen that he believes this right is conditional. Accordingly, liberal peoples may use economic coercion or armed force to intervene in outlaw states if they have tolerated or committed genocide or other crimes against humanity among their people.[57] Importantly, such humanitarian intervention does not properly count as aggression; instead, it counts as third-party self-defense insofar as the population that suffers at the hands of outlaw states is unable to mount its own defense.[58]

In short, for Rawls, a liberal commitment to international law and principles is not a commitment to nonviolence in the nonideal contexts of international politics. However, it is a commitment to establish and maintain the basic conditions of justice as fairness among liberal and illiberal peoples alike. Indeed, liberal peoples must extend toleration to non-liberal peoples who are not aggressive or hostile to fundamental human rights, for the practice of toleration is a requirement of liberalism.[59] However, this requirement is not absolute for Rawls, since on his view outlaw states cannot be tolerated and therefore must be contained or defeated.[60] Let us now turn to examine more specifically Rawls' account of the outlaw state, the role which Rawls describes that liberal nuclearism plays in the former's containment or defeat, and the paradox which Rawls did not foresee concerning the role of liberal nuclearism in the production of liberal outlaw states.

THE OUTLAW STATE AND THE ADVOCACY OF LIBERAL NUCLEARISM

Rawls begins his contrast of liberal peoples and outlaw states by recognizing that IR scholars take the state as a basic unit of analysis and as such are not concerned with "peoples" in their ontology of International Relations.[61] For Rawls's purpose in theorizing a Law of Peoples, states generally are governments that may or may not have their citizen's best interests at heart. In contrast to liberal peoples, which limit their basic interests to what is reasonable (e.g., the security of the citizenry, preservation of liberal practices) and whose relations comport with justice as fairness, states are best understood as rational actors that are primarily concerned with gaining and keeping power. The rationality of states is thus associated with the means-ends calculation of egoism, where preferences are taken as given (and therefore not analyzed according to a notion of moral value) and the means are appropriate insofar as they realize the specified preferences.

After making this distinction between states and peoples, it is interesting that Rawls' concern for states in the IR sense dissipates altogether.[62] Hereafter, his concern is to define the "outlaw state" as that regime which refuses "to comply with a reasonable Law of Peoples"[63] and thus is an irreconcilable foe of liberal and domestic international orders. David Reidy argues that Rawls's account of outlaw states recalls Kant's discussion on unjust enemies.[64] For Kant, the unjust enemy is that state whose expressed will

> reveals a maxim by which, if it were made a universal rule, any condition of peace among nations would be impossible and, instead, a state of nature would be perpetuated. Violation of public contracts is an expression of this sort.[65]

Thus, if we assume that, by some unforeseen circumstance, an outlaw state has become a signatory of a Law of Peoples, it is nonetheless an unjust enemy insofar as its subsequent refusal to comply with a reasonable Law of Peoples counts as an instance of violating public contracts. Or, an outlaw state might refuse to become a signatory of a Law of Peoples and thus signal their hostility to a Society of Peoples. For Reidy, this means that outlaw states are sociopathic: that is, they fail "to possess the capacity to be reasonable to the requisite minimum degree," which is necessary for the security of a Society of Peoples.[66]

We might then conclude that, for Rawls, an outlaw state's character requires noncompliance with a reasonable Law of Peoples. Domestically, outlaw states employ the rule of force, terror, and brutality. They routinely disregard fundamental human rights, such as freedom from slavery or serfdom, liberty of conscience, and freedom from (the threat of) genocide or ethnic cleansing.[67] Internationally, outlaw states are not concerned with honoring treaty commitments. They commit aggression in order to expand their power or influence, and they are not concerned with honoring *jus in bello* constraints during wartime. And if they succeed in conquering a foreign territory, they impose their unjust practices on the subjugated populations. Like Nazi Germany, outlaw states are inclined to pursue weapons of mass destruction that they can use on liberal peoples.[68] Given their nature, outlaw states recognize "no possibility at all of a political relationship with [their] enemies."[69] Rather, these enemies are to be "cowed by terror and brutality, and ruled by force."[70] If outlaw states truly orient their foreign policies according to the maxim of the rule of (nuclear) force, then for Rawls liberal societies are left with no choice except to meet (threats of) force with equal or greater (threats of) force.

It is at this point in Rawls's account that he imports Michael Walzer's argument for liberal (or liberal communitarian) nuclearism, which is presumably grounded on the fifth principle of a Law of Peoples.[71] It is important to note that Rawls does not advance his own moral argument for liberal nuclearism; he merely posits it as a necessary and symmetrical response to the (nuclear) threats of outlaw states. Indeed, Rawls's mention of liberal nuclear deterrence is almost made in passing. This section's analysis must then assume that an examination of Walzer's defense of nuclear deterrence reflects Rawls's own views.

From a general "morality of war" perspective reviewed in chapters 2 and 3, the threat of armed force is permissible if armed force in self-defense is permissible. The converse assumption also seems to hold: that it is wrong to threaten to do something that is wrong to do. This converse assumption has been called the Wrongful Intentions Principle.[72] In this vein, Rawls accepts that the right of national self-defense is conditioned by the *jus in bello* principles of noncombatant immunity and proportionality,[73] which I take as the underlying content of the seventh principle of the Law of Peoples related above. According to Samuel Freeman, this principle expresses the idea that "within war, the human rights of enemy noncombatants are to be respected; noncombatants are not be targeted for attack, and measures should be taken to protect them and their property from injury."[74] This is to say, the right of national self-defense does not permit a state to do anything other than use defensive force in a precise manner to defeat an aggressor's military and, if necessary, its government.

At this point, the question is raised concerning the moral appropriateness of a liberal people's (threat to) use of nuclear force to defend against outlaw state (nuclear) aggression. Walzer's view on this is clear: nuclear aggression is a grave evil, and on the Wrongful Intentions Principle the threat of it is also evil. Even so, the immorality of the threat is not essentially in its utterance; rather, it is in the need to follow through with the threat if deterrence fails. Paradoxically, for Walzer the imperative to prevent nuclear war and its effects introduces the possibility that an otherwise immoral act might become morally necessary. The hope is that nuclear deterrent threats will not fail and nuclear aggression and its horrors are thereby prevented without firing a single shot. In a nuclear-armed world, though, deterrence failure and nuclear reprisal are persistent possibilities. Walzer's agonizing conclusion is that

> deterrence is a way of coping with [the permanence of the supreme emergency[75] of the prospect of mutual nuclear annihilation], and though it is a bad way, there may well be no other that is practical in a world of sovereign and suspicious states. We threaten evil in order to not do it, and the doing of it would be so terrible that the threat seems in comparison to be morally defensible.[76]

Walzer's remark on coping with the prospects of nuclear aggression is that nuclear deterrence is the nonideal provision for preventing the overthrow of the ultimate moral goods that are embodied by national communities. For Rawls, it is the nonideal provision for preventing the destruction of liberal peoples and, if established, a nascent Society of Peoples. The imminent prospect of the near-complete or complete destruction of a national people is what Walzer means by the term "supreme emergency condition." And the crux of his argument turns on the fine distinction between the intent to prevent evil (which leads to the willingness to pose a nuclear deterrent threat) and the intent to not do evil (which would happen if the threat had to be carried out). Without this distinction, the moral justification of nuclear deterrence is incoherent.

On my view, it seems that Walzer's palpable difficulty in wringing out a moral justification of nuclear deterrence is a function of his attempt to reconcile the right of self-defense with the *jus in bello* imperatives of noncombatant immunity and proportional defense against state adversaries that disregard the latter imperatives. One reason is that Walzer's conception of the supreme emergency condition remains sufficiently vague and susceptible to abuse by state leaders.[77] Walzer tries to clarify by stating that Winston Churchill might have correctly described Britain's situation relative to the Nazi Germany in 1940–1941 as a supreme emergency when the fate of Great Britain was in the balance during the German bombing campaign. However, their fate was no longer in the balance by mid-1942 and certainly not afterward when the Allies achieved the upper hand in the war. Hence, Churchill's defense of British obliteration bombing could no longer avoid the requirements of *jus in bello* imperatives because "the supreme emergency [for Britain] passed long before the British bombing reached its crescendo."[78]

If these reflections on Walzer's struggle with the competing demands of *jus ad bellum* and *jus in bello* principles are applied to Rawls's LP, the opposites that need reconciliation are the rule of international law and the rule of nuclear force. Rawls seems to think that the defensive use of armed force does not necessarily mean that the rule of law has been subverted by the rule of force. For, under conditions where outlaw states as unjust enemies refuse to comply with the Law of Peoples or the international law based on it, the (threat of the) use of force is necessary for the preservation of the rule of law. It follows on this account that, if outlaw states (seek to) possess nuclear weapons, a liberal nuclear deterrent could not be considered inconsistent with the rule of international law. The paradoxical outcome suggested by Rawls is that, for liberal order, the rule of force must guarantee the rule of law.

Even so, it is not clear that Rawls believes nuclear deterrent threats against outlaw states should ever be carried out. Rawls indicates that the question of the morality of nuclear war is independent of the morality of nuclear threats, and like Walzer he condemns the United States' use of atomic weapons on Japan in World War II.[79] For Walzer, nuclear war explodes just war theory; likewise, nuclear war explodes the Law of Peoples and the liberal commitment to international law and the moral order that these weapons are tasked to secure.[80] It now seems that Rawls (and Walzer) have overlooked at least one necessary consideration that is part and parcel of deterrence theory: namely, the conditions under which a credible nuclear deterrent threat is possible. It is this consideration that reveals the moral incoherence at the heart of Rawls' commitment to liberal nuclearism and the corresponding and irreversible crisis for liberal ontological security.

THE ONTOLOGICAL AND MORAL INCOHERENCE OF LIBERAL NUCLEARISM

Rawls's liberal peoples exercise reasonableness over and above rationality insofar as "just liberal peoples limit their basic interests as required by the reasonable."[81] As previous sections have noted, the reasonable limits on national interests are defined by the Law of Peoples. Rawls's distinction between liberal peoples and outlaw states rests on the former's exercise of reasonableness or, as Reidy puts it, the latter's sociopathic lack of reasonableness in the exercise of interest.[82] However, Rawls does not deny that outlaw states are rational actors. Thus, the question of liberal nuclearism focused against outlaw states raises the question of this posture's reasonableness as opposed to its rationality. Rawls in LP implicitly recognizes that credible nuclear deterrent threats require adequate nuclear retaliatory capabilities; however, there is nothing in his account that recognizes that credible threats also require the determination to carry them out if deterrence fails.[83] As the previous two chapters revealed, the moral debate during and after the Cold War on the im/permissibility of nuclear deterrence was concerned largely with the intentions of state leaders in making such threats. It was shown that Kantian-inspired commentators rejected nuclear deterrence because it is morally wrong to form and indefinitely sustain murderous intentions.[84] In contrast, many IR realists and rational-choice theorists (or rationalists) denied that the primary intention in nuclear deterrence is the commission of mass murder. Rather, it is the prevention of great power nuclear war—or any war that might lead to a great power nuclear war—and thus it didn't follow that nuclear threats as such were immoral.[85] But, even if nuclear threats were murderous, the importance of the survival of liberal democracy against outlaw state aggression seemed to offer an overriding justification, namely, that the preservation of ultimate liberal goods might paradoxically require the resort to immoral action. The next few paragraphs test this inference by contrasting the rational actor logic

of nuclear deterrence with the core features of Rawlsian political liberalism
to show that the latter must succumb to the former, and then it supports this
finding with reference to the history of nuclear strategy from the writings of
Daniel Ellsberg.

The Logic of Nuclear Deterrence: A Contrast with
Rawls's Kantian Liberalism

In the same passage cited above where Rawls advocates for liberal nuclearism, he intimates that outlaw states can be deterred by nuclear threats. This is
presumably due to the outlaw state's rational calculation that survival under
containment is better than the risk of regime elimination in retaliation for
military aggression. Unfortunately for his account, Rawls appears to not recognize that any government's choice to adopt nuclear deterrence is also a fundamentally rational calculation. If an outlaw state's hostility toward liberal
peoples is Hobbesian (and perhaps sociopathic), then it must be recognized
that the reciprocal hostility of liberal peoples toward outlaw states leading
to the adoption of nuclear deterrence is also Hobbesian. In short, there is no
significant difference between Hobbes' description of sovereign states armed
to the teeth in the mode of mutual deterrence and Rawls' description of an
international order composed of mutually hostile and nuclear-armed liberal
societies and outlaw states.

Moreover, although Rawls insists that liberal peoples are concerned to
observe reasonable limits in the exercise of self-defense, and that this concern
is evidence of their commitment to the rule of international law, their reliance
on nuclear deterrence is ultimately inconsistent with these liberal commitments. To put it a bit differently, nuclear deterrence is not reasonable. One
inconsistency relates to whether a limited self-defense defined by *jus in bello*
constraints would be adequate against outlaw state aggression. It is difficult
to read Rawls as answering this question in the affirmative. Rawls's construction of the outlaw state depicts them as so incorrigible that only the most
severe or decisive measures of self-defense will work. Against this posture
toward outlaw states is the liberal commitment to the human rights of citizens in enemy states, namely, that noncombatants are not to be targeted nor
unduly harmed in any other way. And yet, the commitment to nuclear deterrence makes possible (and perhaps likely) the prospect of nuclear reprisal
on the innocent citizens of outlaw states. Indeed, the supposed effectiveness
of nuclear deterrence in the Cold War was predicated upon the knowledge
that large urban areas were held as nuclear hostages, and that reprisal strikes
would impose unacceptable damages on both countries if aggression were
to occur.[86] If the post–Cold War threat from outlaw states requires the same
basic nuclear deterrent strategy, then it is difficult to avoid the conclusion

that the nuclear threats that must be made against outlaw state populations or assets exceed *jus in bello* constraints, and this constitutes a contradiction between liberal identity narratives of reasonableness, even if it might be consistent with a "great power" and "security provider" identity conceptions.

To address this inconsistency, Rawls implicitly relies on Walzer's moral justification of nuclear deterrence as predicated on the morality of intentions. Walzer's remark quoted earlier emphasized that we "threaten evil *in order to not do it*" as if the intention that qualifies the threat counts as an adequate moral balancer once it is clear that "the doing of [this evil] would be so terrible that the threat seems in comparison to be morally defensible." Hence, the Rawlsian account seems committed to the idea that the utterance of nuclear threats against outlaw states is permissible if (1) outlaw states are deterrable and deterred, and if (2) the threat does not entail murderous intentions. If this analysis is correct, then, like the U.S. Catholic Bishops and J. Bryan Hehir's position on nuclear deterrence during the Cold War, the Rawlsian strategy appears to support a strategy of nuclear bluffing, all the while hoping that outlaw states will not have the temerity to call it.[87] It would then follow that Rawls not only overlooked the general requirement that deterrence threats must be credible but specifically that such threats must be credible to *outlaw states*.

One important implication of this analysis is that the Rawlsian hope that nuclear bluffing can effectively deter outlaw states without the intentional use of nuclear force is the thin reed on which the reasonableness of liberal security policy on this matter rests. However, assuming that outlaw states are rational actors, it must be expected that they understand the strategy of nuclear bluffing as an attempt to maintain the identity of liberal reasonableness. The failure to persuade outlaw states of the credibility of nuclear threats makes deterrence failure more likely than not, even if other regional or international dynamics provide adequate cause for outlaw states to refrain from acting against the vital interests of liberal peoples. The only assured means of convincing outlaw states of the credibility of liberal nuclear threats is to rationally intend to carry out the threat if deterrence fails. And yet, this transforms a reasonable liberal democratic people into a Hobbesian rational actor without due regard for *jus in bello* constraints, thereby inducing a crisis of liberal ontological security.

If we contrast Rawls' advocacy of nuclear deterrence with the Kantian account of perpetual peace on which it is supposed to rest, we find additional reasons for thinking that liberal nuclearism is ultimately not reasonable. With all the appropriate caveats regarding applications of Kant to the Cold War and post–Cold War eras, Rawls's liberal nuclearism sits uneasily with Kant's third preliminary article regarding the eventual need of states to abolish standing armies.[88] Kant recognized that states might need standing armies

to deter rising or immediate security threats during periods of regional instability. However, Kant believed it more likely that state leaders would take
advantage of standing armies to fight wars of aggression. Kant thus realized
that any state's maintenance of standing armies imposes on its neighbors and
rivals an unremitting security dilemma that, in turn, incentivizes arms races
and increases the chances of war. Rawls' liberal nuclearism amounts to a
standing nuclear-armed force that incentivizes the post–Cold War resurgence
of illiberal nuclearism (e.g., Russia, China) and the limited spread of nuclear
weapons to other countries across the world. The more entrenched is liberal
nuclearism, the less inclined liberal peoples will be to roll back these forces
as Kant prescribed for standing armies.

The implications of a liberal people's assignment of Kant's third preliminary article to "irrelevance" include the suspension of the duty to
establish an adequate bond of trust with one's rivals necessary to forge an
international order around perpetual peace. This is the point of Kant's sixth
preliminary article, which states, "No state at war with another shall allow
itself such acts of hostility as would have to make mutual trust impossible
during a future peace."[89] Kant's examples of proscribed acts during wartime—for example, assassination, poisoning—were, of course, indicative
of the limits of eighteenth-century military and technological capabilities.
Suitably updated for the nuclear age, we might read Kant as permitting
the inclusion of nuclear threats as proscribed hostile actions. For, although
threats of nuclear reprisal are not covert acts of violence and betrayal like
assassination and poisoning, they communicate the kind of hostility that
makes mutual trust difficult, if not practically impossible, to cultivate or
sustain. I believe Rawls would acknowledge that nuclear deterrence is not
a trust-building strategy, but perhaps Rawls did not see trust building as
important if there is no possibility of peace between liberal societies and
outlaw states. Yet, outlaw states are not irrational: if their interests dictate
avoidance of aggression, then Kant's sixth principle on maintaining the
possibilities for trust among enemies remains applicable. This is additional
reason to believe that the general effect of the Rawlsian advocacy of liberal
nuclearism is a crisis of liberal ontological security and the corresponding
condition of moral incoherence.

In general, the condition of moral incoherence, beyond mundane levels of
moral hypocrisy, indicates a people's inability to resolve the moral dilemmas
of domestic or security policy. For liberal peoples, moral incoherence reveals
the inability to resolve moral dilemmas in favor of liberal principles such as
human rights and the rule of law that are essential to liberal identity. As such,
liberal moral incoherence is a condition of moral inadequacy, and it is evidence for the judgment that a liberal people have failed to be "who they are"
as a liberal people. By abandoning habits of reasonability, a liberal people

(independent of any other identity conception they might hold, such as great power or security provider[90]) is transformed into an outlaw state.

This moral incoherence of liberal nuclearism is described effectively by Ken Booth and Nicholas J. Wheeler's analysis of ideological fundamentalism in Western security policy. They argue that, during the Cold War, U.S. foreign policy toward the former Soviet Union was informed significantly by ideological convictions about the moral evil of communism. Rather than a measured assessment of Soviet foreign policy in relation to their historic national interests, the dominant wing of U.S. foreign policy advisors believed that the very character of the Soviet Union constituted a fundamental challenge to American democracy and values.[91] According to this Manichean logic, diplomacy or accommodation could not be seen as anything other than signals of weakness or appeasement. This logic fostered a corresponding tendency to exaggerate Soviet military capabilities, to misread Soviet intentions and signals as always implying aggression, and to underestimate the degree to which Soviet leaders felt threatened by hostile U.S. statements and military build-ups.[92] Nothing could reassure U.S. leaders that Soviet policies might rather have been a defensive response to a perceived Western tendency for expansion.[93] It was this ideological fear that drove the United States to undertake an unprecedented expansion of nuclear arms in the early Cold War and settle on the unreasonable and genocidal policy of mutually assured destruction.

Rawls's depiction of outlaw states is comparable to U.S. leaders' depictions of the former Soviet Union and to post–Cold War U.S. administrations' depictions of "rogue state actors." Rawls's invocation of Nazi Germany as an exemplar outlaw state recalls the incessant comparisons by the Bush 41, Clinton, and Bush 43 administrations' depictions of Iraq's Saddam Hussein and Syria's Bashar al-Assad with Adolf Hitler.[94] Rawls's depiction of outlaw states' treatment of dissident minorities recalls the Nazi treatment of Jews, and the characterization that outlaw states will not observe international agreements recalls the Nazi's rejection of the terms of the League of Nations and its betrayal of Stalin when Hitler violated their nonaggression pact. The ideologically charged elements in Rawls's account mirrors the securitizing discourse against Iraq after September 11, 2001, and it is this securitizing discourse that opens the space for liberal nuclear deterrence.

As a result, the ideological fear of the "outlaw" in Rawls's account enervates liberal reasonableness and a willingness to observe *jus in bello* constraints against enemies that are depicted as incapable of humane action. Indeed, it takes the transformation of liberal societies beyond that of Hobbesian rational actors and into that of Manichean enemies in which peace is incomprehensible and for whom total war must be pursued after the first shot is taken. Furthermore, it naturalizes and normalizes the sustained deployment of humankind's most indiscriminate and destructive weapons as the only instruments that can possibly contain the otherwise uncontainable axes of evil. Finally, in the event of deterrence failure, the ideological fear of the "outlaw" will assuage the liberal conscience that carrying out the "unintended" nuclear deterrent threat was the only reasonable option left to them—that is, the outlaw states made it to where liberal peoples had "no choice" but to launch indiscriminate nuclear strikes. And the moment that the liberal people's conscience is falsely assuaged by intentionally doing the unintended, the moral incoherence of nuclear liberalism will have become fully realized.

CONCLUSION: U.S. NUCLEAR STRATEGY AND LIBERAL INCOHERENCE

This chapter has concentrated on developing a theoretical account concerning the ontological and moral incoherence of liberal nuclearism in response to Rawls's account of political liberalism and his subsequent defense of nuclear deterrence against outlaw states. The main thrust of the argument has been that liberal ontological and moral incoherence is produced by liberal nuclearism. Thus, while the commitment to nuclear defense and deterrence might rationally satisfy a great power's national or collective security requirements or perhaps even desire for enhanced international status,[95] it is contradictory to the liberal cosmopolitan commitment to human rights and to the rule of international law. It follows that liberal nuclearism induces ontological dissonance in the liberal state, and tragically the history of the post–World War II era strongly suggests that liberalism in the liberal state eventually succumbs to what Elaine Scarry has called "nuclear monarchy" or what Daniel Deudney has called "nuclear despotism."[96]

It is important to end this chapter on a more concrete note—that is, with a brief discussion of some historical and contemporary evidence of liberal ontological and moral incoherence. To do this, the remainder of the chapter reviews the historical account of U.S. Cold War nuclear war planning by Daniel Ellsberg along with his moral reflections on his participation in this effort as a RAND consultant to the Defense Department from the late 1950s to the 1970s.

Daniel Ellsberg and the Erosion of Liberal Constraints on the Chief Executive

Daniel Ellsberg's recent book-length account of his role in U.S. nuclear strategic planning revealed a number of startling facts and findings that are consistent with this chapter's analysis.[97] In his first chapter, Ellsberg relates that his ninth-grade social science teacher in 1944 introduced the sociological concept of "cultural lag" defined as the dynamic where technological progress outpaces moral progress. Afterward, his teacher assigned the task of reporting on information then publicly available on the process of constructing of a uranium-235 bomb. Their next task was to determine if the use of a uranium-235 bomb would be good or bad for the world. According to Ellsberg, he and his classmates unanimously responded in the negative, citing inconsistencies with liberal values such as respect for human rights and the preservation of the environment. After the U.S. atomic attack against Hiroshima one year later, Ellsberg recalled thinking about that attack

> without the strongly biased positive associations that accompanied [American's] first awareness of it in August 1945: that it was "our" weapon, an instrument of American democracy, developed to deter a Nazi bomb, a war-winning weapon and a necessary one . . . to have ended [World War II] without a costly invasion of Japan.[98]

It is notable that Ellsberg's liberal education produced a measure of critical thinking that enabled him and his classmates in 1944 to correctly condemn a hypothetical Nazi use of atomic weapons and then apply that same reasoning to the U.S. ideological justification of the atomic bombing of Japan.

Even so, Ellsberg could not avoid internalizing the American predisposition to uncritically associate Stalin with Hitler and Soviet Communism with Nazism. As Ellsberg puts it, "This decade of ideological immersion as a Cold Warrior was a necessary part of what prepared me for the next decade of work as a government consultant and official on national security."[99] A close companion to Ellsberg's ideological immersion was his mastery of rational decision theory and its application to command and control elements of U.S. nuclear retaliatory forces by senior military officers and the president of the United States (POTUS). Once it was ideologically determined that just cause "permitted" nuclear retaliation against any Soviet invasion or nuclear first-strike, then the main questions turned to conundrums in command and control authority and routines.[100]

One conundrum concerned how the nuclear command staff, including POTUS, should deal with ambiguous early warning signals. The answer was to launch the long-range bombers with nuclear bombs, have them meet at a prearranged staging area, and maintain a circular flight pattern until they received orders to proceed to target or return to base. This answer was called the "fail safe." Ellsberg's analysis as a RAND consultant identified several nagging problems with this solution. First, in the absence of a positive order to proceed to target, the pilots could not determine whether command had implicitly ordered their return to base or whether command had been destroyed, the latter of which meant that the pilots should proceed to target.[101]

A second problem was revealed when Ellsberg learned that the "fail safe" was practiced by the long-range bombers of the Strategic Air Command but not by the short- and medium-ranged bombers at forward-positioned air bases in South Korea and the rest of the Pacific. These latter bombers were slotted to carry 1.1 megaton gravity bombs, but they never practiced flying to designated staging areas, circling to await and execute their order, and then flying home to base. Thus, the "fail safe" was not universally implemented across the air component of the U.S. nuclear triad, making it more likely that a false positive radar or radio signal might provoke a catastrophic U.S. nuclear strike.[102]

A second and more disturbing example concerned the secret pre-delegation of authority to use nuclear weapons by President Eisenhower to naval and air force officers at various levels of command in the Pacific theater.[103] Specifically, Ellsberg learned that Admiral Harry D. Felt, the nuclear control officer at Commander-in-Chief-Pacific (CINCPAC), had received presidential-level authority in writing to initiate nuclear war if communications between Hawaii and Washington, D.C., had been cut off. In turn, Felt had delegated the same authority to officers below him. These acts of pre-delegation of nuclear launch authority "contravened and superseded" the official guidance offered by the U.S. government's own nuclear war plan, which reserved to POTUS the sole authority to order nuclear strikes.[104] Interestingly, Ellsberg's shock at this discovery failed to provoke his recollection that Congress retained the sole constitutional authority to declare war, which meant that no one could correctly assume the constitutionality of any POTUS order of a nuclear (first) strike. In other words, the U.S. nuclear war plan has clearly excluded congressional input from what is arguably the most consequential and existential of all U.S. foreign policies, effectively transforming POTUS into a nuclear monarch or despot.[105] Applied to the feature of constitutionalism in Rawls's conception of political liberalism, the U.S. nuclear war plan and secret policy of pre-delegation of nuclear launch authority from POTUS to command and field officers have been actively subversive of congressional authority and oversight of U.S. military force for the better part of the last seven decades.

A third and even more disturbing example from the viewpoint of liberal values was Ellsberg's discovery that the U.S. nuclear war plan mandated that any nuclear attack against the Soviet Union would also include nuclear attacks against Chinese military and urban targets.[106] Indeed, by 1960, the United States had targeted every city in the Soviet Union and China with a population of 25,000 or more persons.[107] Planners expected that nuclear war with Moscow and Beijing would involve over 100 million Soviet deaths (which amounted to over 50 percent of this population) and over 300 million Chinese deaths (or about 50 percent of that population).[108] According to Ellsberg, only the Marine Corps commandant, General David M. Shoup, objected that the wholesale murder of Russians and Chinese was not part of the American way.[109] Unfortunately, Shoup's view was in the vast minority, and the U.S. nuclear war plan remained virtually unchanged for the remainder of the Cold War, despite some prominent rhetorical revisions emphasizing "flexible response."[110] The upshot is that the elected and appointed officers of the U.S. nuclear command and control structure had little or no intention to spare innocent Soviet or Chinese citizens; to the contrary, their nuclear war plan intentionally targeted innocents and thereby contradicted key liberal values of noncombatant immunity. Any Rawlsian or Walzerian resort to benign intentions in the construction and implementation of U.S. nuclear strategy is belied by the latter's history.

Finally, Ellsberg learned two other disturbing facts about the U.S. nuclear war plan as it was extended into the post–Cold War era. First, the primary U.S. security objective during the Cold War had been to execute a "launch-on-warning" preemptive nuclear first-strike against Soviet and Chinese (and later, Middle Eastern) targets in response to any (immediately looming) conventional aggression perceived as directly threatening U.S. vital interests, and then use the remainder of U.S. nuclear forces to deter Soviet or Chinese nuclear retaliation.[111] This is disturbing for Rawlsian liberalism precisely because *jus ad bellum* and *jus in bello* principles of proportionality consistent with its human rights conceptions are overridden by the deliberate plan to respond disproportionately to (the prospect of) conventional aggression.

Second, Ellsberg learned that the madman theory has been the official stance of nuclear war and deterrence strategy since the Truman administration.[112] It is this stance that explains why the United States since the end of World War II has issued more than twenty-five nuclear threats and actively considered using nuclear weapons in diverse attempts at coercive diplomacy.[113] One might even argue that the presidency of Donald J. Trump is the clearest example to date of the embodiment of the madman theory. At the same time, it is clear that the United States' reliance on the madman theory is a key indicator of ontological and moral incoherence. As Ellsberg puts it,

> Yet what seems to me beyond question is that any social system (not only ours) that has created and maintained a Doomsday Machine and has put a trigger to it, including the first-use of nuclear weapons, in the hands of one human being— anyone, not just this man, still worse in the hands of an unknown number of persons—*is in core aspects mad.* Ours is such a system. We are in the grip of institutionalized madness.[114]

And it is this madness that makes each individual, people, and state within the reach of a nuclear-tipped missile or bomb hostage to institutionalized rational sociopathy. For Ellsberg, this madness has a blinding effect: "Few Americans are aware of the extent to which the United States and NATO first-use doctrine has long isolated the United States and its close allies morally and politically from world opinion."[115] Indeed, this lack of awareness is the product of ideological isolation, and the inability of a so-called liberal people to think or empathize in appropriately cosmopolitan ways.

In the end, when it comes to the questions of ontological security and moral coherence, it might be that a Hobbesian rational actor, or Rawlsian outlaw state, can accept the grave risks accompanying a reliance on nuclear defense and deterrence capabilities in the name of survival without suffering a rupture in national identity conceptions or the incoherence of a rationalist or consequentialist morality. On the other hand, it is difficult, if not impossible, to conceive that a liberal democratic people or state, especially one that belongs to a liberal international order, can do the same without subverting their liberal identity and contradicting their basic moral commitments to justice as fairness.

NOTES

1. This chapter is an adaptation of Doyle II (2015).
2. Rawls (1999). See also Rawls (1999 (1971)) and Rawls (1996).
3. Walzer (2015 (1977)).
4. The term "outlaw states" arises in Rawls's discussion of states, which will not accommodate to a law of peoples. See Rawls (1999, 80–95).
5. See, for example, Croft (2012); Mitzen (2006a, b); Pratt (2017); Steele (2008); Zarakol (2010).
6. Mitzen (2006b, 342).
7. See, for example, Art (1980); Booth and Wheeler (2008, especially chapters 1–3); Mearsheimer (2014); Morgan (2003); Quinlan (1984 (2009)); Roberts (2016); Schelling (1980 (1960)); Walt (1985).
8. Mitzen (2006b).
9. Steele (2008, 49–75).
10. This explanation comes from Jack Donnelly, quoted by Steele (2008, 95).
11. Steele (2008, 94–113).
12. Lupovici (2012).
13. Lupovici (2012, 816–19).
14. Lupovici (2012, 821–31).
15. Cronin (2001).
16. See, for example, Mearsheimer's definition of hegemony: Mearsheimer (2014, 40–42).
17. Cronin (2001, 105).
18. Cronin (2001, 113).
19. Cronin (2001, 123).
20. See Doyle II (2015b, 27–28).
21. Callendar (1955).
22. Heuser (1998, 97).

23. Steele (2008, 49–75).

24. Steele (2008, 1–25). See also Crawford (2002, esp. 114).

25. Steele (2008, 114–47).

26. See, for example, Kant (1795, 1996d). For relevant commentary, see, for example, Hoffe (2006); Ion (2012); Nardin (2017); Rawls (1999).

27. Freeman (2003, 1–3).

28. Rawls (1999 (1971)); Rawls (1996).

29. Rawls (1999 (1971), 42, 72).

30. Rawls (1999, 132–40).

31. Rawls (1996, 227–30).

32. Rawls (1999 (1971), 194).

33. Rawls (1999 (1971), 197).

34. Sklar (1989).

35. Sklar (1989, 27).

36. Sklar (1989, 23–24).

37. Sklar (1989, 29).

38. Sklar (1989, 29).

39. Kant (1795, 1996d); Wenar (2013).

40. Rawls (1999, 10).

41. Rawls (1999 (1971), 331); Rawls (1999, 4).

42. Rawls (1999, 7).

43. Doyle (1983). See also Russett et al. (1993); Russett (1998).

44. Rawls (1999, 8).

45. Rawls (1999, 9).

46. Russett (1998).

47. Kant (1795, 1996d, §8:354). Clearly, this is also a Hobbesian assumption, and this marks an "IR realist" moment in Kant's analysis.

48. Kant (1795, 1996d, §8:350–357); Rawls (1999, 3).

49. Rawls (1996, 24).

50. Rawls (1999, 37).

51. Weiss et al. (2017, 157–212).

52. Rawls (1999, 37).

53. For a more complete discussion of the ethical dilemmas of nuclear foreign policy, see Doyle II (2013).

54. Rawls (1999, 37).

55. Weiss et al. (2017, 70–122).

56. Rawls (1999, 91).

57. Rawls (1999, 81).

58. Freeman (2003, 47).

59. Rawls (1999, 16–19).

60. Rawls (1999, 81).

61. Rawls (1999, 27–29).

62. Commentators have debated the value of Rawls's distinction between states and peoples for his overall project. See, for example, Buchanan (2000, 698–701); Brock (2010, 89). Although I agree that this distinction is problematic, I am more

concerned with the contrasting character descriptions between liberal societies and outlaw states insofar as nuclear deterrence is capable of securing the former from the latter.

63. Rawls (1999, 5, 90).
64. Reidy (2004, 315, note 17); Kant (1996c, 486–87, paras. 59–60).
65. Kant (1996c, 487, para. 60).
66. Reidy (2004, 315, note 17).
67. Rawls (1999, 78–79).
68. Rawls (1999, 9).
69. Rawls (1999, 99).
70. Rawls (1999, 99).
71. Rawls (1999, 95 fn. 8); Walzer (2015 (1977), chapter 17).
72. Walzer (2015 (1977), 271).
73. Rawls (1999, 94–101).
74. Freeman (2003, 47).
75. Walzer (2015 (1977), 250–67).
76. Walzer (2015 (1977), 273).
77. Shue (2004, 149–52).
78. Walzer (2015 (1977), 259).
79. Rawls (1999, 98–101); Walzer (2015 (1977), 262–67).
80. Walzer (2015 (1977), 281).
81. Rawls (1999, 29).
82. Reidy (2004, 315, note 17).
83. For a small sample of the discussion on nuclear deterrence and its requirements, see, for example, Freedman (2004); Gauthier (1984); Kahn (1985); Kavka (1978); Mearsheimer (2014); Morgan (2003); Quinlan (1981 (2009)); Quinlan (1984 (2009)); Roberts (2016); Schelling (1980 (1960)); Schelling (1966).
84. See, for example, Dummett (1984).
85. See, for example, Kavka (1978).
86. Morgan (2003, 30). See also Ellsberg (2017).
87. See chapter 2. Also see Hehir (1975, 1986); National Conference of Catholic Bishops (1983).
88. Kant (1795, 1996d, §8:344–345).
89. Kant (1795, 1996d, §8:346).
90. See Lupovici (2012). Also, for conflicts among great power and liberal hegemonic identities, see Cronin (2001).
91. Booth and Wheeler (2008, 65–67).
92. Booth and Wheeler (2008, 51–58).
93. Booth and Wheeler (2008, 69).
94. See, for example, Golding (2013); Jentleson (2010, 263)
95. Sagan (1996–97).
96. Deudney (2007, 244–64); Scarry (2014).
97. Ellsberg (2017).
98. Ellsberg (2017, 27).
99. Ellsberg (2017, 30–31).

100. Ellsberg (2017, chapter 2).
101. Ellsberg (2017, 43–46).
102. Ellsberg (2017, 46–59).
103. Ellsberg (2017, chapter 3).
104. Ellsberg (2017, 72).
105. See Deudney (2007); Scarry (2014).
106. Ellsberg (2017, 83–89).
107. Ellsberg (2017, 99).
108. Ellsberg (2017, 102).
109. Ellsberg (2017, 104).
110. Ellsberg (2017, 318). For another detailed account as to how "revisions" to the U.S. nuclear war plan were not implemented at the level of operations, see Gavin (2012).
111. Ellsberg (2017, 12, 318).
112. Ellsberg (2017, 14).
113. Ellsberg (2017, 322). For an extended analysis of nuclear coercive diplomacy and its failures, see Sechser and Fuhrmann (2017).
114. Ellsberg (2017, 332).
115. Ellsberg (2017, 325).

Chapter 5

A Morally Responsible Nuclear Disarmament[1]

The book's second and third chapters critically mapped the nuclear ethical debate that began in the Cold War era and which continued into the present post–Cold War times. These chapters argued that the debate did not progress beyond a core disagreement over which moral principles were ultimately decisive on the appropriateness and moral consequences of nuclear defense and deterrence policies. On one side were commentators whose moral justification of nuclear defense and deterrence policies are based on the just war theoretic right of national self-defense and the requirement of IR realism and rationalism (understood as compatible moral perspectives) to follow through with nuclear threats if deterrence fails. On the other side were commentators who argued that nuclear defense and deterrence are morally abhorrent. Not only did nuclear use (except in unrealistically rare cases) violate the well-established just war principle of noncombatant immunity, but it necessarily put at risk the survival of humanity itself. Such horrific and apocalyptic prospects made even the practice of nuclear threats (i.e., nuclear deterrence) morally abhorrent.

The latter part of chapter 3 explored theoretical means to reconcile these competing viewpoints by introducing a conception of common security as a new anchoring moral principle for nuclear ethics. With this common security principle, the mutual insecurities of nuclear-armed adversaries can be mitigated or resolved under the maxim of each enjoying "security with" the other(s). If the moral imperative of common security were to effectively organize great power politics, then nuclear defense and deterrence policies would no longer be necessary and a corresponding opportunity would be opened for nuclear abolition to proceed without objection.

One of the major obstacles to realizing common security is the great difficulty of each nuclear-armed adversary to overcome entrenched and ideologically charged images of its historic rivals as hated and untrustworthy enemies.[2] These kinds of entrenched images have helped sustain the commitment to nuclear defense and deterrence since the end of World War II. Accordingly, chapter 4's analysis turned to critically examine the political and moral effect of nuclear defense and deterrence policies on liberal democratic institutions and identities in the nuclear-armed democracies. It argued that "liberal nuclearism" is inconsistent with liberal constitutionalism. This is to say, nuclear democracies have witnessed a concentration of war-making power in the chief executive instead of maintaining the shared nature of war powers between the executive and legislative branches. Moreover, liberal nuclearism is also inconsistent with the liberal commitment to the preservation of human rights. In the exercise of nuclear deterrence, nuclear-armed liberal powers cannot avoid cultivating murderous intentions against innocent people dwelling in their enemies' borders. And, should deterrence fail, nuclear-armed liberal powers, such as the United States, have committed themselves to nuclear reprisal attacks against adversaries' urban centers as well as their military assets.[3] Finally, liberal nuclearism is also inconsistent with the liberal commitment to the rule of international law. For, as chapter 4 and this chapter argue, nuclear-armed liberal states' emphasis on all other states' compliance with nonproliferation obligations comes at the expense of their own compliance with nuclear disarmament obligations as related in the 1968 Nuclear Nonproliferation Treaty (NPT). It concluded that if liberal nuclearism produces a rupture in liberal institutions, and even liberal identity itself, then the only morally consistent stance for liberal democracies to embrace is nuclear disarmament.

Accordingly, this fifth chapter focuses on how nuclear disarmament must also be realized in a morally responsible manner. First, it reviews the policy debate on nuclear disarmament motivated by the Humanitarian Imperative to Abolish Nuclear Weapons (HINW) and the contrasting viewpoint advanced by the nuclear-weapon states (NWS) known as "conditions-focused" disarmament. Afterward, it critically examines the moral challenges and dilemmas of achieving nuclear abolition. It concludes by proposing a sketch framework of a morally responsible approach to nuclear abolition informed by the common security imperative.

THE POLICY DEBATE: IMMEDIATE VERSUS
CONDITIONS-FOCUSED NUCLEAR DISARMAMENT

Article VIII of the NPT established the practice of a quinquennial review conference (RevCon) for states-parties to assess the progress of treaty implementation on its missions of nonproliferation, disarmament, and the cultivation of peaceful nuclear energy. Beyond assessing the treaty regime's progress, the RevCons are also charged with setting goals for the next five-year period.[4] After the end of the Cold War, the regime's NWS and non-nuclear-weapon states (NNWS) reaffirmed their commitments to nuclear disarmament in the Final Reports of the 1995, 2000, and 2010 NPT RevCons. During the 2017 and 2018 Preparatory Committee (PrepCom) sessions for the 2020 NPT Rev-Con, the United States, Russia, and China once again reiterated their support for nuclear abolition as did several NNWS, such as Austria, Egypt, Iran, and Ireland.[5] Even so, sharp disagreements continue to roil policy discussions on the best strategy of achieving nuclear disarmament; and these disagreements have led some antinuclear advocates to doubt the sincerity of NWS's rhetorical fidelity to this ultimate objective.

For the purposes of rethinking nuclear ethics for the contemporary era (see chapter 1, introduction), it is important to critically review the competing arguments for regarding nuclear disarmament as an intrinsic good as opposed to merely an instrumental value, that is, one that may be realized if the relevant international security conditions are propitious. The following critical review thus provides a descriptive groundwork for the ethical analysis in the latter parts of the chapter on a morally responsible nuclear disarmament.

The Humanitarian Imperative to Abolish Nuclear Weapons

On August 6, 2014, the sixty-ninth anniversary of the atomic bombing of Hiroshima by the United States, the International Federation of Red Cross and Red Crescent Societies (IFRC) and the International Committee of the Red Cross (ICRC) released a joint statement titled *Remembering Hiroshima: Nuclear Disarmament Is a Humanitarian Imperative.*[6] It recalled a series of ICRC Council of Delegates resolutions in 2011, which leading abolitionist NNWS cited in their 2018 PrepCom remarks.[7] These ICRC resolutions expressed deep moral concern

> about the destructive power of nuclear weapons, the unspeakable human suffering they cause, the difficulty of controlling their effects in space and time, the threat they pose to the environment and to future generations, and the risks of escalation they create.[8]

The joint IFRC/ICRC statement thus expressed a long-standing concern of the global antinuclear movement that nuclear war or other large-scale nuclear accidents inherently constitute a grave moral evil for humankind.[9] Accordingly, it appealed to all states to ensure that nuclear weapons are never used again and to pursue negotiations toward a complete and irreversible elimination of nuclear weapons based on existing commitments and international obligations.

Other global civil society actors joined with the IFRC/ICRC joint statement, including Nobel Peace Prize Laureate Archbishop Desmond Tutu from South Africa. In one publication posted on the International Campaign to Abolish Nuclear Weapons (ICAN) website, Tutu raised what he believed to be the central and most stubborn question of the contemporary period: why the N5 continue to state their need for nuclear weapons. His answer invoked what he saw as Cold War inertia and a stubborn attachment to the threat of brute force to assert the primacy of some states over others.[10] For Tutu, neither reason satisfies the requirement of genuine military or moral necessity.[11] Rather, they affirm that, as former United Nations General Secretary Ban Ki-Moon stated, "there are no right hands for wrong weapons."[12] For Tutu, the NPT's apartheid-like distinction between nuclear haves and have-nots reflected a fundamental injustice that only universal and irreversible nuclear abolition could redress. Understanding the political difficulties of nuclear abolition, he called for an irrepressible groundswell of popular opposition against the N5: "By stigmatizing the bomb—as well as those who possess it—we can build tremendous pressure for disarmament."[13]

Archbishop Tutu's call to stigmatize nuclear weapons is a necessary first step in a complex process of constructing an international regime prohibiting their possession.[14] Such stigmatization efforts have faced steady and mostly effective opposition from the N5 and their allies. As reported in previous chapters, several scholars and policy experts have argued that nuclear deterrence has played the largest role in preserving great power peace since the end of World War II. To abolish nuclear weapons, on this view, would risk the end of this era of peace. In contrast, the antinuclear movement and abolitionist NNWS contend that the belief that nuclear deterrence has kept the peace is deeply flawed inasmuch as it overlooks other explanations and is blind to the increasing likelihood of nuclear deterrence failure. Given that nuclear-armed democracies are some of the HINW's key opponents, it became increasingly clear that a determined and irresistible antinuclear global civil society effort to change hearts and minds on nuclear defense and deterrence was required.[15]

The Disarmament Push: From the 13 Points to the 2017 TPNW

The indefinite renewal of the NPT at the 1995 RevCon was predicated largely on an agreement among N5 and NNWS that the end of the Cold War provided an unprecedent opportunity for Article VI disarmament requirements to be effectively implemented.[16] During the 1998 PrepCom meetings for the 2000 NPT RevCon, abolitionist NNWS expressed significant dissatisfaction over the pace of Article VI progress. As a result, eight NNWS formed the New Agenda Coalition (NAC) which took two years to agree on a list of interim measures that, if undertaken, would count as satisfying the Article VI requirements.[17]

Two years later, the 2000 NPT RevCon Final Report listed these measures as the "13 Points."[18] The complete list of these points is as follows:

1. An immediate and unconditional commitment to a Comprehensive Test-Ban Treaty (CTBT);
2. A verifiable moratorium on all nuclear testing until the CTBT's entry into force;
3. An immediate effort within the Conference on Disarmament to bring into force a treaty on banning the production of fissile materials for nuclear explosive devices in a reliable and verifiable manner, otherwise known as the Fissile Materials Control Treaty (FMCT);
4. An immediate effort to establish the mandate for nuclear disarmament within the Conference on Disarmament;
5. A commitment by all states to applying a principle of irreversibility on nuclear disarmament;
6. An "unequivocal undertaking by the nuclear-weapon States to accomplish the total elimination of their nuclear arsenals leading to nuclear disarmament, to which all States parties are committed under Article VI";
7. An immediate undertaking to advance the Strategic Arms Reduction Treaties between the United States and Russia, and the strengthening of the Anti-Ballistic Missile Treaty which had been in force since the Cold War period;
8. The completion and implementation of the Trilateral Initiative between the USA, Russian Federation, and the IAEA;
9. The taking of concrete steps by all NWS toward nuclear disarmament in a way that promotes international stability and security, such as

   a. Unilateral nuclear arms reductions
   b. Increased transparency in the same
   c. Continued reductions of tactical nuclear weapons stocks
   d. De-alerting of nuclear weapons
   e. Diminishing the role of nuclear weapons in national security doctrines
   f. Engagement by all NWS in good faith negotiations toward nuclear disarmament

10. The placement by all NWS of fissile material no longer required for military purposes under IAEA verification protocols;
11. Reaffirmation by all NWS of the ultimate objective of nuclear abolition;
12. Regular reports by all NPT states parties on the progress in implementing Article VI; and
13. The further development of verification capabilities that will ensure compliance by all to their NPT obligations.

Understood in their entirety, the 13 Points translate Article VI's general language on nuclear disarmament into specific measures. They also express mistrust and hope about the disarmament process. For instance, points 4, 6, 9, and 11 separately demand nuclear abolition. This redundancy suggests the NAC's uncertainty over the N5's trustworthiness on this matter, which was produced originally by the U.S. Senate's refusal to ratify the CTBT that President Clinton had signed in 1996.[19] Nonetheless, the NAC retained hope that incremental progress would continue and that nuclear disarmament would be realized sooner than later.

Unfortunately, effective implementation by the N5 of the 13 Points was soon undermined by the United States' response to the September 11, 2001, terrorist attacks on New York, Washington, D.C., and Pennsylvania. One part of the reaction to the 9/11 attacks was found in the 2002 U.S. National Security Strategy, which elevated the role of nuclear weapons in its defense and deterrence postures.[20] At the same time, the United States invaded Afghanistan to root out terrorist safe havens and shortly thereafter invaded Iraq to remove Saddam Hussein's alleged nuclear weapons program. As a result, by the 2005 NPT RevCon, the United States did not agree to any language that reconfirmed its commitments to the 13 Points. Rather, it concentrated on strengthening nonproliferation policies and initiating other measures to prevent nuclear terrorism. In the end, the 2005 RevCon failed to produce a final consensus report largely as a result of the standoff between the abolitionist NNWS and the Bush administration that, along with several U.S. allies, were fearful of what seemed to be a new and dangerous nuclear threat environment.[21]

Consequently, the 2008 election of Barack Obama to the U.S. presidency was notable for his public recommitment to the path of nuclear disarmament as defined by the 13 Points.[22] Nuclear abolitionists' renewed hope motivated efforts to further operationalize the 13 Points by means of a 64-Point Action Plan, which the N5 accepted unanimously for inclusion into the 2010 NPT RevCon Final Report. Although many of these action plan statements were reiterations of the 13 Points, others were new. These included (but were not limited to) the following:

1. The N5's recognition of the legitimate interest of the NNWS in the former's reduction of the operational status of their nuclear forces;
2. The N5 to finalize commitments to provide negative security assurances to NNWS;
3. The establishment of new nuclear-weapon-free-zones, especially in the Middle East;
4. A greater concentration on cultivating cooperation among all members of the United Nations and other regional and international organizations related to nuclear disarmament issues; and

5. The implementation of programs of disarmament and nonproliferation education so as to advance the goals of a complete, irreversible, and verifiable nuclear abolition.[23]

As these five action statements reveal, the abolitionist NNWS urgently lobbied the N5 to avoid any further delay in honoring their Article VI commitments. Thus, while the abolitionist NNWS "could not realistically expect the Action Plan to be [fully] implemented by 2015, . . . they certainly did not adopt it with the intention of observing slow movement for another half a century."[24] Even so, the strong contrary preference of the N5 was for a slower step-by-step disarmament process. Unsurprisingly, abolitionists interpreted the N5's response as a delaying and appeasement strategy, which in turn strengthened their resolve to press for nuclear disarmament without delay.

Ultimately, the 2010 RevCon Final Report marked the third and final moment of agreement between the abolitionist NNWS and N5 on the nature of Article VI commitments. By 2013, it was obvious that progress on the 2010 Action Plan had stalled. In that same year, the first of three international conferences on the HINW was convened in Norway and the others were held, respectively, in Mexico (2014) and Austria (2015). In the Austria conference, a Humanitarian Pledge was endorsed by 127 states, and a subsequent endorsement of the HINW was delivered by former Austrian foreign affairs minister, Sebastian Kurz, to the 2015 NPT RevCon.[25]

Unfortunately, by the end of the 2015 RevCon, abolitionist NNWS' hopes were completely dashed as they confronted a set of dismal realities. First, the progress on CTBT entry-into-force had stalled by the refusal of eight of forty-four "Annex 2" states to ratify the treaty.[26] These included India, Israel, Pakistan, Iran, North Korea, and even Egypt which was a NAC member. However, abolitionist NNWS were more concerned that the United States and China had not ratified the CTBT given their affirmations of 2010 Action Plan. Second, the N5 had not worked further to establish a mandate on nuclear abolition within the Conference on Disarmament. Third, the N5 refused to apply a principle of irreversibility to nuclear arms reductions or to nuclear disarmament generally. Finally, and perhaps most tellingly, the N5 blocked any inclusion of the language or narrative of the HINW into the 2015 RevCon Final Report.[27] In response to these developments, the abolitionist NNWS began a separate process of negotiations that eventually led to the adoption of the Treaty on the Prohibition of Nuclear Weapons (TPNW) in the United Nations General Assembly on July 7, 2017.[28]

However, prior to the TPNW vote in the General Assembly, the 2017 NPT PrepCom meetings were convened during the latter part of April and the first part of May. Anticipating the N5's continuing objections against the intensified disarmament demand, Ireland's Permanent Representative to the United Nations, Patricia O'Brien, spoke for the NAC and argued forcefully that

> the NAC considers that the global security situation cannot justify the lack of progress on nuclear disarmament. On the contrary, it reinforces the need for urgent action. What is lacking is not the propitious conditions, but the political will and determination.[29]

In a subsequent statement, O'Brien added,

> A number of NPT States Parties believe that nuclear disarmament can only be achieved in a gradual manner, if and when national and international security conditions so permit. It is regrettable that this approach has so far objectively produced little result in practice.[30]

O'Brien strongly suggested that the primary delaying tactic of the N5 had been to seek for the resolution of new nuclear challenges or security dilemmas prior to the full implementation of the 2010 RevCon disarmament commitments. Indeed, she implied that the N5 thinking on this issue was backward. Instead of conditioning nuclear disarmament upon the realization of international and regional stability, O'Brien contended that "international stability and undiminished security for all is the strongest argument in favor of accelerated progress towards achieving a nuclear-weapon-free world."[31]

Echoing O'Brien's rhetoric, the non-aligned states-parties to the NPT proposed for the 2020 RevCon a three-phase action plan that, over the course of fifteen years, could provide a time-certain for the complete elimination of nuclear weapons.[32] From 2020 to 2025, the first phase would involve a dual effort to initiate an effective disarmament process. One effort would involve the immediate implementation of the 64-Point Action Plan. The other part would be the commencement and conclusion of negotiations leading to a comprehensive convention that would ban and destroy nuclear weapons under a "single integrated multilateral comprehensive verification system." From 2025 to 2030, the second phase would involve accelerated action to bring this comprehensive convention on nuclear abolition into force. Afterward, states would act in concert to establish the convention's verification system, declare nuclear stocks, inventory nuclear arsenals, separate warheads from delivery vehicles, store these warheads securely for future destruction, and remix warhead fissile materials for peaceful energy usage. From 2030 to 2035, the requirements of the comprehensive convention on nuclear weapons would be completely satisfied. All nuclear weapons would be eliminated in an irreversible and verifiable manner. All military nuclear facilities would be converted for peaceful uses. And all nuclear materials, equipment, and technologies would be placed under IAEA safeguards and oversight.

Consistent with the NAC and Non-Aligned states' PrepCom proposals, the abolitionist NNWS sought universality in the TPNW negotiation
process leading up to the July 2017 meeting in the General Assembly. To
their disappointment, but not their surprise, the N5 and almost all states
covered by U.S. nuclear deterrence guarantees (i.e., the "umbrella states")
boycotted the TPNW negotiations. The Netherlands was the only umbrella
state to attend the negotiations.[33] Some commentators doubted that the
TPNW could succeed in actually banning nuclear weapons without the
involvement of the N5, but its advocates nevertheless argued for the treaty's
political utility to disrupt the nuclear *status quo* and compel the N5 to face
their nuclear policy dilemmas.[34] When the votes were finally tallied in the
General Assembly on July 7, 122 states out of the 124 attendees voted in
favor of the TPNW.[35] Thereafter, the TPNW's ratification process began
and, as of late-2019, it had 34 states-parties out of the fifty required for it
to come into force.[36]

The 2018 NPT PrepCom was convened nine months after the July 2017
General Assembly vote on the TPNW. Abolitionist NNWS representatives'
statements reaffirmed the HINW and intensified pressure for immediate action
on nuclear disarmament. Some moved beyond reiteration of familiar abolitionist points to emphasize the need for alternative security conceptions—for
example, human security over and above narrow national security—as well
as emphasizing new research on the unequal and more devastating effects of
nuclear detonation effects on women.[37] These and other abolitionist NNWS
reaffirmed the NPT as a pillar of international security as well as emphasized
that the TPNW was a facilitating instrument for the success of Article VI
disarmament objectives.

Looking toward the 2020 NPT RevCon and beyond, significant uncertainty
remains concerning the prospect of disarmament realization. Arguably, the
legitimacy and efficacy of the TPNW depends in part on its coming into force
sooner than later. It took the NPT two years to come into force after passing
the General Assembly in 1968. As I see it, if the TPNW fails to come into
force by early 2021, then chances increase that it will suffer the same fate as
the moribund CTBT or, at the very minimum, fail to achieve the degree of
stigmatization of nuclear weapons its authors seek. Two points of concern
are important to raise. First, only three of the NAC states so far have become
TPNW states-parties—Mexico, South Africa, and New Zealand—and it is
important for a number of reasons that the remaining NAC states ratify the
treaty without further delay. Second, the entire roster of current TPNW states-
parties seems unable at this moment to influence the behavior of the N5 as
their power relations are currently configured. Indeed, even if the TPNW
came into force by virtue of gaining fifty or more smaller or weaker states, it
is unlikely to make a significant difference if the N5 and their umbrella allies
can effectively counter-stigmatize the nuclear ban movement.[38]

The Conditions-Focused Disarmament Discourse

The most recent competing viewpoint to that expressed by the abolitionist NNWS has been the "conditions-focused" disarmament discourse. (The terminology of the discourse has recently changed to "environment-focused.") The United States has been the foremost proponent of this view,[39] although Russia, China, and the umbrella NNWS have echoed its main points.[40] The conditions-focused disarmament argument starts with the claim that nuclear abolition is an important goal for the international community. Even so, nuclear disarmament is feasible only if it can be completed in a safe, verifiable, and sustainable process. In turn, its safety and sustainability are dependent on an adequate set of resolutions to the regional and global security dilemmas that promote trust among adversaries.[41]

For instance, Chinese and Russian statements during the 2017 and 2018 PrepComs identified the United States' policy of first-use of nuclear weapons as one of the main obstacles to their own progress on nuclear disarmament. China's statements reiterated its commitment to no-first-use and to nuclear use only as a last resort in response to nuclear aggression by the United States.[42] Beijing's distrust of Washington is not only rooted in the latter's insistence on retaining a first-use policy but also in its nuclear sharing policies with key umbrella NNWS allies. Beijing believes such nuclear sharing contravenes NPT Article I prohibitions on transferring nuclear weapons to NNWS. Moreover, Beijing sees the forward deployment of U.S. ballistic missile defense systems to South Korea and elsewhere as intensifying regional and international insecurities. Ultimately, Beijing argues that each of these U.S. policies would have to be reversed before progress on nuclear abolition could be safely undertaken.[43]

Russian 2017 PrepCom statements echo and amplify China's emphasis on the special responsibility of the United States in the nuclear disarmament process.[44] Additionally, Russia rejected any notion that it had failed to make progress on its Article VI obligations. It reported reductions of its nuclear arsenal by approximately 80 percent from Cold War levels, much of that in conjunction with the United States. Nonetheless, this progress had been "overshadowed by systematic violation by the United States of the [Intermediate Nuclear Forces] INF Treaty."[45] Specifically, Moscow believed that the installation of U.S. ballistic missile defense systems in eastern European NATO states counted as INF violations, and it expressed agreement with Beijing that U.S.-NATO nuclear sharing arrangements violated Articles I and II of the NPT. Hence, Moscow and Beijing agreed that any progress on nuclear disarmament must follow suitable changes in U.S. nuclear policy.

The United States' conditions-focused nuclear disarmament discourse rests on a theory advanced by former president Ronald Reagan that nations "do not mistrust each other because they are armed; they are armed because they mistrust each other."[46] For U.S. officials, Reagan's claim is consistent with the NPT Preamble that seemed to cast the "easing of international tensions and the strengthening of trust between states" as a disarmament prerequisite.[47] Accordingly, the United States asserted that Russia and China's resurgence as great power competitors, along with escalating nuclear threats from North Korea and Iran, made any immediate or unilateral nuclear disarmament overly risky and perhaps disastrous. It cited Russian aggression into Georgia in 2008 and Ukraine in 2013–14 in support of this view, along with the annexation of Crimea; Iran's violation of the IAEA Safeguard Agreement in the early 2000s and its continued pursuit of the nuclear threshold; North Korea's withdrawal from the NPT in the early 2000s and their 2006–19 program of nuclear and ballistic missile tests; and the movement by Russia, China, and others to expand or modernize their nuclear arsenals.[48] On the basis of these new threats, the United States feared that Russia, China, or other rivals might take advantage of any conscientious disarmament effort to cheat and then attempt a policy of nuclear blackmail.[49]

As a whole, N5 representatives concurred that the abolitionist NNWS put the cart before the horse by claiming that nuclear disarmament would resolve nuclear adversaries' security dilemmas.[50] According to Christopher A. Ford, an assistant secretary of state in the Bureau of International Security and Nonproliferation,

> The TPNW is perhaps the paradigmatic example of an approach to disarmament that is willfully blind to the challenges and complexities of the real world in which decisions about national and international security and nuclear disarmament actually occur. And in its deliberately provocative and divisive approach—drawing moralistic lines between stakeholders in the international community, for instance, slandering and demonizing those who disagree with its crusade, urging key states to take positions they understand to be gravely detrimental to those states' national security, and ignoring imperatives tied to deterring aggression and maintaining alliance relationships that underpin peace and security in critical areas of the world—the TPNW model represents precisely the *opposite* of what one would want from a serious effort to ease tension and strengthen trust in ways conducive to disarmament progress.[51]

Ford's argument seems to preemptively dismiss the abolitionist NNWS's emphasis on human security. Instead, he charges that TPNW advocates are "willfully blind" by choosing to ignore that nuclear and non-nuclear adversaries are compelled to act according to their national security interests. He rejects the abolitionists' claim that the United States is guilty of divisiveness for not following the TPNW "moral crusade." Rather, he implies that the abolitionist NNWS have introduced an unnecessarily divisive element to regional and international politics by alienating the very states with whom they must secure cooperation.

If Ford and others are right on these points, it would follow that privileging the 13 Points, the 64-Point Action Plan, or the requirements specified by the TPNW is imprudent. Rather, the morally and politically responsible course of action would be to realize nuclear disarmament by resolving the key security dilemmas among nuclear armed rivals and suspected nuclear aspirants (such as Iran). This is not to say that proponents of conditions-focused disarmament perceive the resolution of key security dilemmas and all disarmament measures as mutually exclusive. For example, the question of disarmament verification has become the key overlapping area of interest among some abolitionist NNWS and the N5. During the 2017 NPT PrepCom, Andre Haspels of the Netherlands remarked that "if we want a world without nuclear weapons, we need an ironclad verification mechanism that at the same time prevents the transfer of proliferation-sensitive weapons-related information."[52] Haspels' use of "ironclad verification mechanism" was meant to address the uncertainties about the prospects of nuclear cheating once an arms control or disarmament agreement had come into force. If the security dilemmas of NWS adversaries were partially linked to fears of cheating, then agreeing on suitable and strict verification mechanisms seemed to satisfy both the conditions-focused disarmament concern of the N5 as well as the demands of abolitionist NNWS to honor the commitments to the 13 Points.

Haspels' remark might be taken as a reminder to all NPT states-parties of their shared interests affirmed seven years previously. Soon after the conclusion of the 2010 NPT RevCon, the Non-Proliferation and Disarmament Initiative (NPDI) was formed. Composed of twelve member states (Australia, Canada, Chile, Germany, Japan, Mexico, the Netherlands, Nigeria, the Philippines, Poland, Turkey, and the United Arab Emirates), the NPDI focused on increasing transparency among all NPT members states—and especially the N5—as well as the reinvigoration of disarmament discussions in the Conference on Disarmament.[53] By forming such coalitions, the abolitionist NNWS (such as Mexico, Chile, and Nigeria) and key umbrella NNWS (such as Germany, Japan, the Netherlands) sought to make concerted progress on both the disarmament agenda and the mitigation of NWS's security dilemmas. During the 2018 NPT PrepCom, the NPDI submitted a working paper urging the universal adoption of the IAEA Additional Protocols designed to "prevent the diversion of nuclear materials from peaceful uses to nuclear weapons or other nuclear explosive devices."[54] The NPDI praised the 148 states that had signed the Additional Protocol, although it noted that 16 of those signatories had not yet submitted the relevant ratification documents. Iran was one of those signatories, and states such as North Korea and Pakistan, which had not yet signed the Additional Protocol, continued to fuel the kinds of uncertainties motivating the N5 to retain their nuclear arsenals. In a post-disarmament world, the Additional Protocol's function consistent with Actions 28–30 from the 2010 NPT

RevCon Final Report would be to maintain absolute and verifiable prohibitions on the diversion of nuclear materials to non-peaceful activities.[55] In order to succeed at this latter mission, the NPDI acknowledged that the technical details of the safeguards should evolve over time, and that NPT states-parties should willingly accept revisions to the protocols as they become necessary.[56]

The NPDI's statements concerning the imperative of verification were echoed by European Union (EU) representatives. Like the NPDI, the EU is comprised of states on both sides of the TPNW divide. On one side is Ireland, a leading NAC member, and on the other side are a number of umbrella NNWS along with France, the EU's foremost NWS. During the 2018 NPT PrepCom, the EU reaffirmed Haspels' 2017 statement by devoting an entire section of their working paper to the idea that effective verification was the key to nuclear disarmament.[57] Moreover, the EU emphasized that the NNWS needed to play an increased role in the establishment of an effective verification process. Thus, the EU working paper counts as a sincere effort to incorporate into a policy agenda key interests from opposed NPT coalitions. This is evident in their contention that

> transparency, irreversibility, and verifiability remain at the core of any disarmament regime, the protection of sensitive and proliferative information, managed access, completeness and correctness of host declarations and safety and security are all essential issues. The right balance must be achieved.[58]

On the one hand, the EU acknowledged the importance of the abolitionist NNWS demand for transparency, irreversibility, and verifiability, each of which were elements of the 13 Points and the 64-Point Action Plan. On the other hand, the EU implicitly recognized that practices of transparency also risked an irresponsible release of sensitive and "proliferative" information concerning any NWS stockpile or capability during a disarmament process. Only when a "right balance" had been achieved between the equally important values at stake could disarmament proceed responsibly. The EU expressed faith in their ability to facilitate and maintain such a balance given their successful sixty-year track record on verification.[59] Additionally, they pointed to the example of the Trilateral Initiative of the IAEA between 1996 and 2002, showing that the United States and the Russian Federation were capable of managing the shared challenges of verification of nuclear reductions without revealing sensitive information.[60] Thus, the EU maintained confidence that continued N5/NNWS collaboration on verification would provide the necessary common ground on which future nuclear disarmament negotiations might proceed.

The NPDI and EU statements on verification to the 2017 and 2018 NPT PrepCom did not, however, convince the abolitionist NNWS that

conditions-focused disarmament approach was enough for meaningful progress on Article VI commitments. Thus, in late 2018, Christopher A. Ford
addressed the Nuclear Threat Initiative, a U.S.-based civil society organization, on the need to build a conditions-focused disarmament discourse
that could ease abolitionists' concerns.[61] During the first year of the Trump
administration, Ford was surprised to learn that Obama administration officials had accomplished little to nothing in policy to advance their rhetorical
commitments to nuclear disarmament. Thereafter, he researched thoroughly
the complete range of options on the question of nuclear disarmament and
decided that the conditions-focused disarmament approach was the most
responsible path forward. For Ford, the pressing question was how to build an
NPT-wide consensus on this approach given the passionate support of nuclear
abolition by so many NNWS.

Ford's answer to his own question recommended a multifaceted public
diplomacy effort. One facet would involve expert roundtables that would
refine or sharpen the theoretical frameworks of national and international
security that bear upon the dilemmas of nuclear disarmament. Another
would involve regional workshops to engage with policy experts and civil
society advocates to apply this refined thinking to the policy level. Finally,
NPT states-parties would seek safe and verifiable ways to enact the interim
disarmament measures identified by the 13 Points. A crucial step in Ford's
proposed effort was the inclusion of people outside the "community of disarmament 'usual suspects.'"[62] Presumably, Ford meant the inclusion of TPNW
advocates to serve the purpose of consensus building. Part of that process
would involve persuading TPNW advocates to avoid "drawing moralistic
lines" and "demonizing those who disagree with [their] crusade" rather than
working on how to "ease tensions and strengthen trust in ways conducive to
disarmament progress."[63] On the other hand, for such consensus building to
succeed, it would also be important for those in Ford's camp to learn if U.S.
criticisms of abolitionist NNWS were grossly mistaken: that is, if U.S. threat
perceptions were exaggerated and if U.S. nuclear defense and deterrence
policy in its current configurations has in fact put humanity's existence at
grave risk.

The foregoing analysis of the NNWS disarmament push suggests that the
Cold War–era stalemate in the debate on the strategic and moral justifiability of nuclear defense and deterrence related in previous chapters has been
extended into the contemporary debate on the political and moral value of
banning nuclear weapons. Given the asymmetries of power among NPT
states-parties, it seems that the TPNW advocates face a pair of vexing questions. First, do they have sufficient political clout to induce an effective moral
stigmatization of nuclear weapons and thereby persuade the conditions-
focused disarmament advocates to abolish nuclear weapons or, at the very

least, make it politically impossible for the N5 to avoid nuclear abolition anymore? And, if the answer to the first challenge is affirmative, are they able to achieve nuclear abolition without violating other binding moral requirements themselves? These questions raise the prospect of a second kind of "conditions-focused" disarmament discourse that draws on the moral imperative of common security advanced in chapter 3.

A Common Security Approach to Nuclear Disarmament

A common security approach to nuclear disarmament must be responsive to the Humanitarian Imperative as well as to the true and vital national security interests of the N5 and umbrella states. A moral imperative of common security (chapter 3) is the anchoring principle of this approach, and it functions similarly to Kant's categorical imperative insofar as it is the criterion against which rules or maxims for national or collective defense and deterrence policies are measured. This approach is comprised of four elements that are discussed in the next few paragraphs.

Civil Society Activism

First and foremost, a common security approach to nuclear disarmament expects that the N5 will not conform their behavior to the 13 Steps, the 2010 Action Plan, or the TPNW until organized, sustained, and sufficiently large domestic and global civil society pressures are mounted. As Lawrence Wittner contended,

> Given the tension between the widespread desire for nuclear disarmament and the national security priorities of the nation-state, nuclear policy usually has proved a rough compromise, unsatisfactory to either the nuclear enthusiast or critic. Often it takes the form of arms control, which regulates or stabilizes the arms race rather than bringing it to an end. . . . What, then, will it take to abolish nuclear weapons? As this study suggests, it will certainly require a vigilant citizenry, supportive of peace and disarmament, groups that will settle for nothing less than banning the Bomb.[64]

Wittner, an avid nuclear disarmament advocate, implicitly acknowledges that the N5 have formed their nuclear policies to confront two dangers: the unacceptable dangers of a world of nuclear-armed states and the unacceptable vulnerability of a nuclear-free world without adequate measures to address their security dilemmas. The resulting stalemate has produced an unhappy compromise position of nuclear restraint.[65] Disarmament advocates assert correctly that a perpetual regime of nuclear restraint is inconsistent with N5 Article VI commitments as defined by the 1995, 2000, and 2010 NPT

RevCon Final Reports. Even so, Wittner correctly argues that the N5 are not likely to act on those commitments absent an organized and sustained pressure from a sufficiently large antinuclear civil society movement. This is to say, the politics of nuclear restraint will continue until domestic public pressures reconfigure public debate and electoral politics in favor of nuclear abolition. Accordingly, for this kind of pressure to truly become irresistible, the antinuclear movement must embrace a dual focus: national efforts to effectively pressure adoption of nuclear abolition and international efforts to strengthen relevant international legal and organizational elements.[66]

Wittner's view echoes the institutionalist analysis of Ethan Nadelmann, who argues that global civil-society pressures to strengthen international legal and organizational elements are necessary for any domestic political effort to produce changes in states' foreign policies.[67] Citing cases of the domestic and international efforts against piracy, privateering, and slave trade, Nadelmann identifies five stages of the evolution of a prohibition regime.[68] First, a targeted activity (e.g., slavery, piracy), which had been regarded as legitimate, is opposed by domestic and international actors whose interests conflict with that of those promoting the targeted activity. As a result, the targeted activity becomes the object of increased regulation and control. This first stage counts as a normative shift for a targeted activity away from unconditioned to conditional acceptance. In relation to nuclear weapons, this first evolutionary stage was relatively short insofar as the shift to regulating nuclear weapons possession and manufacture occurred within the first two decades of the nuclear age—that is, the 1950s and 1960s.[69]

Second, the targeted activity is stigmatized by redefining it as no longer permissible for civilized international society. This stigmatization effort is often led by norm entrepreneurs such as legal experts, religious leaders, or public intellectuals. Nadelmann recalls the role of Christian religious organizations in England and the United States in the eighteenth and nineteenth centuries in the stigmatization of the African slave trade and chattel slavery. Such efforts were successful first in British Parliament and afterward within the northern United States.[70] Likewise, nuclear abolition is unlikely unless the possession and use of nuclear weapons are effectively stigmatized. So far, the antinuclear movement has not succeeded in stigmatizing the possession of nuclear weapons, even if a nuclear use taboo remains effective.[71] If it does succeed, additional steps that Nadelmann's account identifies must be taken to continue the progress toward nuclear abolition.

For Nadelmann, the third stage involves effective state efforts to criminalize the targeted activity by means of an international convention. States might undertake diplomatic initiatives, offer economic inducements, threaten military action, or otherwise push for a formalized prohibition instrument. The fourth stage involves the creation of the relevant prohibition regime and its coming into force. This is the stage that corresponds to the TPNW's expressed prohibition on nuclear weapons; that is, unless it is meant strictly or primarily as a stigmatization instrument. The final stage involves the corresponding decline of the targeted activity to no more than marginal or obscure instances.[72]

Nadelmann's account of the evolution of prohibition regimes explains how the ban on nuclear proliferation emerged and has endured in the NPT regime.[73] In the first stage, the NWS realized their mutual interest in preventing any further spread of nuclear weapons. In the second stage, the NWS and many other NNWS effectively stigmatized new nuclear proliferation. Once this stigmatization had sufficiently diffused, the third and fourth stages witnessed the negotiation, conclusion, and ratification of the NPT. The NPT as an international convention effectively "criminalized" new nuclear proliferation among states-parties, and by the post–Cold War period the NPT regime was nearly universal. The NPT grand bargain included the economic incentives of assured access to peaceful nuclear energy in exchange for accepting an absolute ban on nuclear proliferation. And, as is common knowledge by now, the resort to UN Security Council sanctions as punishment for violating the treaty's nonproliferation provisions has occurred most recently against North Korea and Iran. As a result, the fifth stage has witnessed how the NPT mostly succeeded at preventing the spread of nuclear weapons over the past fifty years.[74]

If Nadelmann and Wittner's analyses are correct, then one might reasonably wonder about the efficacy of using the TPNW, a *prima facie* stage-four measure, to induce stigmatization, a stage-two outcome. Moreover, it cannot be overemphasized that without reaching key constituencies in N5 states, the disarmament norm will not evolve beyond the first stage. Even more worrisome, the failure of stigmatization of nuclear weapons might well exacerbate the erosion of the nuclear taboo.[75]

The Increasing Risk of Nuclear Deterrence Failure

The second element of a common security approach to nuclear abolition involves making an effective case to the NPT and non-NPT NWS (i.e., India, Pakistan, Israel, and North Korea) that nuclear deterrence is bound to fail at some moment, and that catastrophic nuclear reprisals will likely follow any first-use of nuclear weapons.[76] Even key advocates of nuclear deterrence during the Cold War admitted that "it is difficult to believe that [nuclear deterrence] will last forever."[77] Accordingly, as reliance on nuclear deterrence approaches its eighth decade, the growing awareness of eventual deterrence failure is not based merely on the fact of its management and operation by human beings prone to accident, misperception, and miscalculation. It is also that nuclear deterrence policies tend to destabilize regional geopolitical interactions. According to Francis Gavin, during the Cold War U.S. nuclear weapons frequently

> nullified the influence of other, more traditional forms of power, such as conventional forces and economic strength, allowing the [former] Soviet Union to minimize the United States' enormous economic, technological, and even "soft power" advantages. Nuclear weapons also changed military calculations in potentially dangerous ways. It has long been understood that in a nuclear environment, the side that strikes first gains an overwhelming military advantage. This meant that strategies of preemption, and even preventive war, were enormously appealing.[78]

Gavin's two-part observation suggests that nuclear deterrence failure is multifaceted. First, overreliance on nuclear deterrence paradoxically can erode a state's overall deterrence posture, leaving it vulnerable to decreases in overall influence and power. This means that nuclear weapons can empower and disempower simultaneously. Furthermore, Gavin suggests that most U.S. officials' understanding of nuclear deterrence failure is inadequate. That is, more than a failure of the United States to prevent nuclear first-use by North Korea, Russia, or another NWS, it is also the failure of any NWS to prevent the kinds of conventional aggression that can lead to nuclear first-use or even the failure of nuclear deterrence to prevent new nuclear nonproliferation. In the latter case, U.S. nuclear deterrence was as much an effort to prevent key NNWS allies, such as Germany, from acquiring their own nuclear deterrents, which would increase the risks of Russian or Chinese aggression or even cause difficulty for managing relations among NATO allies.[79] Finally, it must be recognized that the failure of any foreign NWS to deter the United States from engaging in military action against their vital national interest counts as deterrence failure. For instance, any preventive strike against North Korea to remove the Kim regime and its nuclear weapons would count as a failure of North Korean nuclear deterrence, with all of the negative consequences for the United States should Kim decide to "use them or lose them."[80]

Without question, nuclear abolition is the most direct method of preventing nuclear deterrence failures; but the challenge is to persuade NWS policymakers who are convinced otherwise and do not adequately understand how their nuclear deterrence postures contribute to the existential insecurities of others. In the absence of effective civil society pressures or abolitionist NNWS diplomacy, it seems that the only way to effectively stigmatize nuclear weapons is for NWS policymakers to experience and (narrowly) survive a catastrophic nuclear accident or war. But, of course, this is the moral tragedy we must prevent in the first place. A common security approach to nuclear abolition must therefore make every effort to enable determined NWS and their allies and other nuclear aspirant states to perceive nuclear weapons as irrelevant and useless for their security interests.

Security Dilemma Sensibility

In this regard, a third element for a common security approach to nuclear abolition is suggested by Ken Booth and Nicholas J. Wheeler's concept of security dilemma sensibility.[81] Booth and Wheeler define the security dilemma as a two-stage strategic predicament ensnaring states involved in security competition.[82] The first stage involves a policymaker's uncertainty about an adversary's intentions or motives regarding its force postures or defense doctrines. This "dilemma of interpretation" is evident in the above-mentioned 2018 NPT PrepCom statements by China over the United States' refusal to make a no-first-use pledge.[83] For the United States, maintaining a nuclear first-use right is meant to dissuade any other NWS first-use of nuclear weapons. For China, however, the U.S. claim to a right of first-use might also signal an underlying intention to engage in nuclear aggression under certain conditions. As a result, China's insecurity is maximized, and, in the words of Michael Walzer, Beijing faces a permanent "supreme emergency condition."[84]

The second stage of the security dilemma involves a policymaker's uncertainty concerning the proper response to their adversary's ambiguous postures or actions. Put schematically, State A might or might not do well by developing or augmenting its own nuclear forces in response to State B. It may be that A might remain safe by not responding tit-for-tat to B's arms build-up, but A might not trust that B's intentions are limited to deterrence. If not, then A might feel it must respond by its own arms build-up, even if by doing so it fuels an arms race.

In this vein, history suggests that state leaders are intensely aware of their own security dilemmas and often lack adequate empathy with those of their adversaries. During the Cold War, President Ronald Reagan expressed surprise that Soviet leaders genuinely believed that U.S. leaders were intent on nuclear aggression against Moscow.[85] Russian leaders seemed similarly unaware that U.S. leaders genuinely feared Soviet nuclear capabilities. Once Reagan and Gorbachev appreciated more fully each other's predicament, they were motivated to take denuclearization steps. Their mutual empathy largely facilitated the historic Intermediate Nuclear Forces (INF) Treaty of 1987.[86] Booth and Wheeler concluded that

> Gorbachev sought . . . to enter into the counter-fear of Western policy-makers by designing a set of policies aimed at fundamentally changing Western threat

perceptions. His unilateral promise to cut those combat forces that most wor-
ried NATO planners . . . was arguably the most dramatic act of reassurance
made during his time in office. The episode is a fascinating example of security dilemma sensibility because it demonstrated that leaders can take steps to
increase their security which, far from decreasing the security of their potential
adversary, actually increases the sense of security felt by both sides.[87]

Two points in Booth and Wheeler's analysis are noteworthy. One is that
security dilemma sensibility involves "entering into the counter-fear" of
one's adversary to understand their threat perceptions. This empathetic act
is the basis of reformulating security policy in order to diminish the other's
fear and insecurity. In other words, security dilemma sensibility anticipates
that one's own commitment to arms reductions will induce the adversary's
commitment to the same, thus leading to a virtuous spiral that might end with
disarmament.

A second point is Booth and Wheeler's emphasis of the role of human
agency in managing the security dilemma. The concept of military or political
necessity, which state leaders often invoke to justify nuclear defense and deterrence policies, implies the absence of effective human agency insofar as actors
"have no choice" but to respond in kind. State B "has no choice" but to respond
to A's arms build-up with its own, or State A would "have no choice" but to
respond to B's nuclear aggression with nuclear reprisal strikes. However, by
"entering into the counter-fear" of an adversary, State A's leader is capable of
understanding how her actions affect State B's perceptions and, consequently,
make it possible for A to imagine alternative courses of action leading to mutually secure outcomes. In the current polarized context of NPT regime politics,
a common security approach to nuclear disarmament would emphasize the
importance of security dilemma sensibility such that abolitionist NNWS could
enter into the counter-fear of the NPT NWS and vice versa. Once mutual empathy has been selected as the dominant strategy, it is more likely that a principled
and practical consensus on disarmament policy can emerge.

Alternative Security Discourses

A fourth and final element of a common security approach to nuclear disarmament is a logical outgrowth of the first three elements: namely, the
adoption of an alternative (i.e., common) security narrative that justifies and
enables the decisions and actions toward disarmament that would otherwise
seem ill-advised. One might wonder if this fourth element ought to precede
the other three. As I see it, it is likely that NWS adversaries will not accept the
common security approach unless civil society pressures are applied and new
knowledge is acquired concerning the increased likelihoods of nuclear deterrence failure and the need for security dilemma sensibility. If this assumption

is correct, then it is necessary to summarize here the common security conception developed more thoroughly in chapter 3.

Rooted in the "security community" thinking of Karl Deutsch[88] and the Palme Commission Report of 1982, the key contention of the common security approach is that

> there can be no hope of victory in a nuclear war, the two sides would be united in suffering and destruction. They can survive only together. *They must achieve security not against the adversary but together with him.* International security must rest on a commitment to joint survival rather than on a threat of mutual destruction.[89]

This is to say, even though conventional national security thinking leads officials to the belief that, ultimately, their state's survival and security is a function of (the threat of) the use of force *against* rivals or enemies, there is no foreseeable scenario in which nuclear defense or offense will enhance national, regional, international, or human security. The quality of life for survivors of nuclear war will be severely compromised, and if escalation dynamics cannot be controlled, then the only foreseeable future will consist of mutual destruction or annihilation. For the Palme Commission, the mutual assuredness of nuclear destruction for the United States and the former Soviet Union removed any content to the notion of victory in a superpower nuclear war. In game-theoretic terms, nuclear war is not a zero-sum game; it is a negative-sum game for all involved. Genuine security, on the other hand, must be a positive-sum game for all involved in the nuclear era.

The case of former Soviet leader Gorbachev is once again illustrative. According to Wittner, Gorbachev was exposed to the new thinking of the Palme Commission and the antinuclear thinking of the Pugwash Conferences.[90] As Gorbachev faced the severe economic and political challenges to the Soviet system in the mid-1980s, he began to understand Soviet security not in terms of a constant stance *against* the United States but *with* it on matters of common concern. And even under the pressures of economic decline, Gorbachev's acceptance of common security enabled a courageous effort of conciliation with the United States, even as Reagan undertook a significant nuclear arms build-up. Eventually, Gorbachev persuaded Reagan of the necessity of eliminating nuclear weapons from the world, but his persuasive success was grounded in the alternative common security approach.[91] Without Gorbachev's willingness to embrace a common security conception, there is little doubt that a new superpower arms race would have continued and that the chances of nuclear war would have been significantly increased.

These four elements of a common security approach to nuclear disarmament— irresistible civil society activism, acknowledgment of the increasing prospects of catastrophic nuclear deterrence failure, the cultivation of security dilemma sensibility, and the embrace of a common security approach to the security dilemma—must be taken as independently necessary and jointly sufficient for the realization of a nuclear-weapon-free world.[92] The history of post–Cold War nuclear politics strongly recommends this conclusion. It begins with the fact that security dilemma sensibility and the embrace of common security by Reagan and Gorbachev at the end of the Cold War lasted only as long as the Reagan-Bush 41 and Gorbachev-Yeltsin era lasted. As chapter 3 details more thoroughly, the action-reaction dynamic between the United States and Russian Federation since the mid-1990s has been characterized by increasing mutual suspicion, lack of empathy, and military postures that entail the intensification of security dilemmas. Both states have withdrawn from important arms control commitments, including the INF Treaty, and have modernized their nuclear forces. Both have elevated the role of nuclear weapons in their national security doctrines. If NWS adversaries are mutually unwilling to enter into each other's counter-fear, then it is also next to impossible to believe that abolitionist NNWS's efforts to stigmatize nuclear weapons will significantly affect NWS thinking and policymaking.

Moral Implications of the Policy Debate

A nuclear ethical analysis of contemporary nuclear politics in the world might start by considering the range of possible outcomes regarding nuclear abolition. One is that efforts to impose an immediate nuclear abolition will fail if the N5 refuse to join the TPNW and continue to resist implementation of any 2010 Action Plan that they believe leaves them unacceptably vulnerable. This outcome would have to count as a significant moral and political failure of the antinuclear movement given the HINW.

Another possible outcome is that efforts to impose an immediate nuclear abolition will succeed because NWS adversaries will have experienced a catastrophic nuclear accident or war that would activate their urgent desire to eliminate any remaining nuclear weapons. Alternately, they might decide that the short-term political costs of maintaining nuclear deterrence are too high if they narrowly avoid a catastrophic nuclear incident. In the former case, the success of nuclear abolition could not overcome the overwhelming moral failure to abolish nuclear weapons prior to nuclear catastrophe, even if it impelled disarmament action thereafter. In the latter case, the pressures of antinuclear civil society groups would appear to have prevailed. However, if that civil society pressure is not accompanied by the combination of the corresponding internalization of the stigmatization of nuclear weapons, the fear of deterrence failure, the practice of security dilemma sensibility, and an embrace of common security, it seems much more likely that the current antinuclear pressures associated with the emergence of the TPNW will diminish as have previous disarmament efforts.[93] With such pressures diminished and the NWS security dilemmas reactivated, it is very likely that nuclear rearmament would occur. The latter case must therefore count as a morally irresponsible and risky outcome.

A fourth possibility is that the N5's preference for a conditions-focused nuclear disarmament leads to a complete and irreversible nuclear disarmament anchored on a transparent and multinational verification regime or, fifth, further instances of modest or significant nuclear reductions but not a complete and irreversible nuclear disarmament. In this latter case, the failure to realize nuclear disarmament amounts to a failure to resolve trust dilemmas among rival NWS and allies.[94] Given the HINW, it is morally irresponsible to fail to resolve trust dilemmas and therefore to fail at realizing nuclear abolition, even if progress continues on nuclear reductions. In the former case, the success of complete and irreversible nuclear abolition must count as a moral and political victory. Unfortunately, it is difficult to imagine the realization of nuclear abolition from a viable verification regime without an effective stigmatization of nuclear weapons and the embrace of common security. It follows on this line of reasoning that the N5 preference for a conditions-focused disarmament along with their faith in nuclear deterrence and a resolute commitment to exclusive national security approaches obviates the prospect of nuclear abolition. It is thus difficult or impossible to conclude that the current conditions-focused disarmament discourse is morally responsible.

A sixth possibility is that a common security approach to conditions-based nuclear disarmament leads to a slow but significant reduction of nuclear weapons down to such low levels as to count as "virtual disarmament" *or*, finally, a complete and irreversible nuclear disarmament consistent with the 13 Points, the 2010 Action Plan, and the TPNW. In this former case, it will be difficult to call "virtual disarmament" a moral victory. Indeed, retention of a small number of nuclear weapons by any state would indicate their lingering faith in nuclear deterrence as well as a stubborn commitment to national security over common security. It follows that even a commitment to a common security approach to conditions-based nuclear disarmament is not guaranteed of a morally responsible outcome. On the other hand, states that have committed to the common security approach have a much better chance of realizing actual nuclear abolition than are states in any other scenario. This is because a commitment to the common security approach is more likely accompanied by significant respect for the HINW. If states are looking to honor the HINW as well as protect their legitimate national interests, then it seems we can conclude that the common security approach is the most responsible in moral terms of the comparable approaches.

MORALLY RESPONSIBLE ENDS AND MEANS: SKETCHING AN ACCOUNT OF A MORALLY RESPONSIBLE NUCLEAR DISARMAMENT

In this final section, the remaining tasks are, first, to address the challenges of the common security approach to nuclear disarmament and, second, to sketch an account of a morally responsible pathway to nuclear disarmament. The sketch of such a pathway counts as one element of the larger effort to rethink nuclear ethics in the twenty-first century, which is the subject of the next and concluding chapter of this book. However, this sketch first must account for the kinds of moral challenges that those committed to a common security approach will likely encounter.

The common security imperative seeks to reconcile the national security and human security imperatives in the effort to protect the world's states and peoples from nuclear catastrophe or extinction. The policy debate between national and human security approaches arises from the incompatibility of these imperatives as perceived by their respective advocates. The former fear that the HINW will be privileged above national security, posing unacceptable risks for national security, and the latter fear that national security will continue to be privileged above the HINW, posing unacceptable risks for human survival. The reconciliatory effort of common security is to simultaneously satisfy national and human security imperatives as state adversaries adopt "security with" instead of "security against" postures.

Nevertheless, it has been argued that a common security approach requires domestic and global civil society activism and public diplomacy by abolitionist states to educate politicians and citizens in N5 and umbrella states concerning the prospects of deterrence failure and the need for security dilemma sensibility. Accordingly, it is expected that the moral dilemmas of common security and the corresponding fears of existential defeat will be most keenly felt during any period where adversaries commit to implementing "security with" arrangements. (Note: although the discussion below is focused on the N5 and umbrella states, it must also include the nuclear-armed states outside the NPT regime: India, Pakistan, Israel, and North Korea).

Addressing the Moral Challenge of Inducing Fear of Nuclear Catastrophe or Human Extinction

A common security approach to nuclear disarmament accepts unconditionally that abolishing nuclear weapons is a humanitarian imperative, although it doubts that a human security approach will persuade the N5 and umbrella allies to undertake what they perceive as an unsafe course of action. Thus, it endorses the mobilization of antinuclear civil society groups to effectively persuade and pressure policymakers in the N5 to undertake without further delay the interim measures identified by the 2000 and 2010 NPT RevCon final reports.[95] We have seen that many antinuclear civil society groups have succeeded in persuading a supermajority of UN member states to embrace the HINW, and ICAN's success in particular was recognized with the 2017 Nobel Peace Prize.[96] The question is, if the antinuclear movement can repeat their successes with the N5 states and their publics, which might then cascade toward success for the remainder of the world's nuclear-armed states, how can that be done?

One action suggested above is the cultivation of a concrete and profound fear of nuclear deterrence failure and its catastrophic effects for states and peoples (hereinafter, nuclear fears). Such nuclear fears have been successfully cultivated among the leaders and members of the antinuclear movement and the abolitionist NNWS.[97] It would therefore seem morally unproblematic to cultivate the same fear in those which possess and defend the possession of nuclear weapons.

However, it is important to not assume that the cultivation of nuclear fears in the name of nuclear abolition is morally unproblematic. Walzer believed nuclear deterrence worked during the Cold War *because* it called up dramatic images of human pain.[98] And, if Westerners feared nuclear death less than the Soviets, it was because they feared the consequences of Soviet conquest even more.[99] Either way, it seemed to Walzer and other like-minded commentators that the cultivation of nuclear fears was morally justified to prevent the greater evil of nuclear war from occurring. In contrast, Western antinuclear activists, such as Jonathan Schell, believed that the experience of nuclear fear inflicted significant psychological and moral harm upon those who knew they were nuclear hostages.[100] Moreover, if nuclear war were to never happen, the trauma associated with the fear of nuclear war's occurrence would remain immoral and unjust. If Schell's view is correct, then cultivating nuclear fears in the name of national or human security was morally problematic and perhaps morally irresponsible.

The moral question of cultivating nuclear fear arises somewhat differently in the post–Cold War era, largely because it has been approximately seventy-five years since the Hiroshima-Nagasaki atomic attacks and over twenty years since the NWS conducted nuclear tests (except for North Korea).[101] The keen sense of nuclear death seems to have significantly diminished except among the increasingly small number of survivors of the 1945 atomic attacks and others who witnessed directly the effects of great power nuclear testing. Hence, it might seem to antinuclear advocates that nuclear fears must be cultivated afresh: for a twenty-first-century nuclear ethics, the question about the moral ir/responsibility of imposing on individuals a new or fresh trauma of nuclear fear in the service of effective antinuclear civil society mobilization.

If examined from a moral consequentialist viewpoint, it seems reasonable that experiencing nuclear fear can lead to participation in the antinuclear movement. Indeed, it explains Ronald Reagan's surprising conversion to nuclear abolitionism.[102] In mid-1983, Reagan saw the film, *The Day After*, which depicted the horrific aftermath of a U.S.-Soviet nuclear war as experienced by survivors in Lawrence, Kansas. Reagan's visceral reaction to this film motivated him to seek a way out of the superpower nuclear dilemma. It became a key anchoring experience for him as he sought a common security relationship with Gorbachev. For the moral consequentialist, Reagan's experience shows that nuclear fear can become the beginning of the wisdom of nuclear abolitionism.

Even so, a deontological moral approach would emphasize that human rights—and our corresponding duties to preserve them—put limits on the exercise of consequentialist-motivated security policy. This cautionary note recalls Kant's categorical imperative regarding the treatment of oneself and others as ends in themselves, and not as mere means.[103] It also recalls Joseph Nye's remark in his *Nuclear Ethics* that "self-defense is a just but limited cause."[104] Accordingly, as the *jus in bello* constraints of noncombatant immunity and proportionality limit the *jus ad bellum* right of national defense, so does the human right of personal security—especially the right to be free from fear—constrain the kinds of actions that are permissible regarding the individual right of security from the dangers of nuclear catastrophe. Hence, just as the noncombatant immunity principle is sufficient to condemn nuclear deterrence and warfighting, the human right of personal security might be thought sufficient to constrain antinuclear activists from cultivating nuclear fears as a motivating force to realize nuclear disarmament.

The force of the deontological side constraints, though, must depend on the corresponding degree to which the looming threat of aggression would result in catastrophic evils. We recall that Walzer claimed that the existence of nuclear weapons alone puts states and peoples in a constant state of "supreme emergency."[105] However, humanity itself is hostage to a constant and immediate nuclear existential threat, and not merely states and peoples as distinct international actors. The force of the deontological side constraints must be weighed against the moral severity of the looming nuclear catastrophe for the human species, and not just this or that state or people. Indeed, this is exactly the point of the Humanitarian Imperative.

Accordingly, it seems appropriate that a common security approach to nuclear disarmament would apply Steven P. Lee's amended Principle of the Morality of Social Institutions (PMSI) to the question of cultivating new or fresh fears of nuclear catastrophe. As related in chapter 2, PMSI states that "social institutions are morally justified only if they achieve their social benefit in a way that does not systematically violate nonconsequentialist rules, such as those of justice and the respect for rights."[106] Applied to the question of nuclear disarmament, PMSI would require that the social benefit of complete and irreversible nuclear disarmament must be accomplished only in a way that does not systematically violate nonconsequentialist rules, such as the right of individuals everywhere to be free from existential fear and harm. However, one might object that PMSI is too strict given the stakes for humanity regarding its nuclear-weapons-induced supreme emergency. It is to this objection that Lee's amended PMSI is introduced. It says that an institution that systematically violates nonconsequentialist rules is nonetheless morally justified only if (1) this institution achieves a sufficiently great social benefit that (2) could not otherwise be achieved without systematic violation of nonconsequentialist rules and for which (3) there are no other alternative institutions to secure this benefit whose violations of nonconsequentialist rules are less severe.[107]

Thus applied, Lee's amended PMSI would permit the systematic cultivation of a new or fresh set of nuclear fears by the antinuclear movement, even though by doing so it violates the individual right of personal security for each person in whom such fear is cultivated. This is because (1a) the social benefit of preserving humanity, or a large segment of it, is sufficiently great, (2a) the history of the nuclear disarmament movement is strong evidence for the role of nuclear fear in realizing major gains in nuclear freezes and reductions, and (3a) no other institutional approach that has been attempted has yet produced acceptable results in realizing nuclear abolition. This list of reasons does not rule out the possibility that a different emotion might have sufficient motivating force. If any such emotion did arise and were as or more effective than the cultivation of nuclear fear, then the latter would no longer enjoy the moral justification of PMSI. However, if the Reagan-Gorbachev case is illustrative, then we might reasonably infer that the exercise of security dilemma sensibility and the embrace of common security among superpower rivals could not have occurred in the absence of their fear of nuclear catastrophe. Recognition of the role of such nuclear fears in other elements of a common security approach to nuclear disarmament strengthens the argument for its moral justifiability on the amended PMSI in the absence of alternative and effective motivating emotions or interests.

Addressing the Moral Challenge of Cultivating Security Dilemma Sensibility

At first glance, the practice of security dilemma sensibility among NWS adversaries, and even among N5 and NNWS in their conflicts over nuclear disarmament, might seem morally uncontroversial. By the mutual exercise of empathy, each state adversary is more likely to make progress toward mitigating or resolving their nuclear-related dilemmas. Moreover, security dilemma sensibility is an approach that Kantian ethics would likely affirm because it is consistent with the requirement of regarding one's adversary as an end and not a means only. Even so, it is important to recognize some significant political difficulties that attend this practice and then determine if our initial and favorable moral judgment needs to be revised.

One difficulty is addressing adversary state officials' sense of fatalism and futility if it seems to them that their conflicts of geopolitical interest are irreconcilable. Of course, this is exactly the case in which security dilemma sensibility must have significant positive effect, since the possibilities of constructing an alternative common security commitment leading to nuclear abolition are not likely to otherwise emerge. Conflicts of geopolitical interests will continue to be perceived as irreconcilable if "security against" assumptions are maintained, and efforts to cultivate empathy will likely be futile in the absence of a commonly held fear of nuclear catastrophe. Hence, it is not that the attempt at forging mutual empathy is morally responsible *per se*. Rather, the political and moral challenge is to see such attempts as indications that a different course of political action is required—namely, the cultivation of nuclear fears.

Another related difficulty is that of sustaining security dilemma sensibility between heads of state in the face of determined domestic and ideologically driven political opposition. For instance, Reagan's foreign policy advisors and many Republican senators opposed his empathetic turn to Gorbachev and then later to the passage of the 1987 INF Treaty.[108] Similarly, the U.S. Senate refused to ratify the CTBT under presidents Clinton and Obama, and they opposed but failed to prevent Obama from arriving at common ground with Iran on the latter's nuclear program in the 2015 Joint Comprehensive Plan of Action (JCPOA).[109] In each of these cases, domestic and ideologically driven opposition believed that U.S. adversaries were not trustworthy because they were communists or Islamic fundamentalists, respectively.[110] They feared that these illiberal regimes would take advantage of the United States' misplaced trust and eventually deter Washington from pursuing its interests or, in the worst case, attack its vital interests directly.

These cases illustrate the limits of security dilemma sensibility if its practice is confined to the level of the heads of state of adversarial countries. Once such heads of state are convinced of the common security conception and have begun the processes of mutual empathy, they must be committed to facilitating and incentivizing the same among their advisors, their allied and opposition-party legislators, their key domestic constituencies, and eventually their respective citizenries. Otherwise, any exercise of security dilemma sensibility at the highest levels of state interaction is likely to be undermined at the lower levels and thereby sets back the cause of sustained security cooperation that, for our purposes, is concentrated on nuclear abolition. Such setbacks are not morally acceptable, and this means that attempts at security dilemma sensibility among rival heads of state are morally irresponsible if undertaken without due regard for the necessary and corresponding commitments at other levels of official interaction.

The recent case of the JCPOA highlights the moral irresponsibility of an abortive process of security dilemma sensibility for the prospects of a successful nuclear abolition. Although President Obama and Secretary of State John Kerry succeeded at establishing a degree of mutual empathy with Iranian President Hassan Rouhani and Foreign Minister Javad Zarif during the negotiations from 2013 to 2015 leading to the adoption of the JCPOA, hardline factions and their key constituencies in both countries maintained an ideologically motivated distrust and opposition to the agreement. In the 2016 general election campaign, the then-candidate Donald J. Trump repeatedly voiced in concert with his Republican allies that the Iran Nuclear Deal was the worst deal that the United States had ever signed.[111] Within eighteen months of his election, President Trump formally withdrew the United States from the JCPOA, and many experts within the arms control and disarmament community have interpreted this withdrawal as a critical foreign policy error.[112] Had successful systematic and persistent efforts been undertaken during (and even before) the Obama administration to diffuse security dilemma sensibility with respect to Iran among key GOP circles, it is likely that the United States would have remained in the JCPOA.

Unfortunately, long-standing opponents of the JCPOA in particular and of nuclear disarmament generally in the United States have made it practically impossible to realize sustained gains in 13 Points and 2010 Action Plan. Thus, the need for effective stigmatization efforts against nuclear weapons and the cultivation of nuclear fears. It is expected that disarmament opponents will mistakenly interpret domestic and international efforts to stigmatize nuclear weapons as "divisive" and "moralistic."[113] Their failure to understand the need for stigmatization is produced by their determined faith in nuclear deterrence and their deliberate lack of empathy for the existential insecurities of the NNWS. They will fail to understand how a deliberate lack of empathy is itself divisive and freighted with the moralism of ideological commitments. They will fail to appreciate the lengths to which the abolitionist NNWS have entered into the counter-fears of the N5 and attempted conciliation through institutions such as the NPDI and Vienna Group of 10. They will fail to appreciate the need for reciprocation of these abolitionist NNWS efforts, that is, unless a small- to modest-sized nuclear first-use event scares them into a commonly held appreciation of the need for nuclear abolition.

In short, a common security approach to nuclear disarmament requires domestic political actors, with the facilitation of each state's chief executive, to courageously abandon ideologically driven postures on security policy and to correspondingly lead public opinion in the formation of appropriate perceptions and orientations to state adversaries. Such perceptions and orientations would take account of the histories of adversary states, including their fears and vulnerabilities. In the end, only an integrated and vertically instituted practice of security dilemma sensibility is likely to enable the kind of horizontally instituted practice of mutual empathy effective at achieving sustained gains in nuclear abolition in a morally responsible manner.

Addressing the Moral Dilemma of State Vulnerability

In the process of pursuing a common security approach to nuclear disarmament, each N5 adversary will likely face difficulties in implementing each of the interim disarmament steps identified in the 2010 Action Plan. In these moments of difficult decision, N5 leaders will feel keenly the increased risks of their state's vulnerability to their nuclear-armed adversaries. The actions taken to implement common security arrangements might have to be made before the cultivation of mutual empathy had significantly advanced, and that means such actions will be undertaken under conditions of significant uncertainty. Yet, even if the process of building mutual empathy had significantly advanced, the perceived vulnerabilities are diminished only after a given NWS' costly signal is reciprocated by its adversaries.[114] In comparison to other actions discussed earlier on a common security approach, the implementation of any of the 2010 Action Plan statements plausibly counts as one of the most difficult and morally dilemmatic of state-level decisions.

Again, the crucial case study of the Reagan-Gorbachev relationship is illustrative. As discussed earlier, Gorbachev had tried to reduce superpower tensions by making concessions to the United States and NATO on a range of nuclear and nonnuclear issues. These concessions were costly signals sent before Reagan had truly appreciated the Soviet fears concerning the prospects of U.S. nuclear aggression. Even so, Gorbachev took the gamble because he did not believe that the United States or NATO would attack if the Soviet Union acted in a nonprovocative way.[115] Gorbachev's domestic opponents complained that the United States would not bypass an opportunity to take advantage of Moscow. This complaint was rational in view of the resistance of U.S. national security advisors against Reagan's abolitionist turn. Fortunately, as related above, Reagan reciprocated Gorbachev's costly signal by expressing willingness to join in nuclear reductions and even by proposing complete nuclear abolition.[116] In this case, then, Gorbachev's costly signal succeeded and later assisted in the efforts at cultivating a mutual empathy.

By contrast, it might be objected that making costly concessions to one's NWS adversary as a result of implementing the key disarmament action statements is morally irresponsible even if there is a foreseeable chance of disarmament success. Several responses might be made to this objection. One is that, if moral responsibility is defined by the consequentialist requirement of success at accomplishing the greater good, then it is mistaken to assess Gorbachev's concessions to the West as morally irresponsible. Although the odds of success might have been (greatly) against him, his success in shifting the U.S.-Soviet relationship away from mutually assured destruction (MAD) and toward cooperative threat reduction (chapter 3) must nonetheless count as a morally responsible act.[117] Secondarily, Gorbachev's success illustrates that misplaced mistrust or distrust is as much of a political and moral problem as misplaced trust.[118] Gorbachev's trust in Reagan was not misplaced, which meant that the act of making Moscow vulnerable to Washington by means of offering concessions was not imprudent. Indeed, had Gorbachev failed to place trust in Reagan, he would have missed the opportunity to change the trajectory of the superpower relationship away from MAD and toward two decades of significant nuclear reductions. To willingly miss such opportunities is to act in a morally irresponsible way.

Finally, to the extent that the current U.S. administration recalls the Reagan-Gorbachev case, they miss its most important lesson: that the trust which Gorbachev and Reagan mutually exercised did not involve the kind of risk calculation that the N5 might associate with morally responsible action in the mode of prudence. Rather, their mutual trust accompanied the formation of a personal bond that transcended the concerns of risk calculation.[119] The two leaders' mutual confidence in each other's personal integrity and the transparency of their intentions were enough to motivate their joint disarmament efforts without concern for the odds of failure. And for Gorbachev and Reagan to not have worked to form that personal bond would have indeed been morally irresponsible.

A Morally Responsible Nuclear Disarmament: Concluding Remarks

This chapter raised the question of the morality of the means of pursuing an ultimate moral good, that is, nuclear abolition. In so doing, it contrasted the policy debate between advocates of an immediate nuclear disarmament versus advocates of a conditions-focused disarmament. This contrast linked the policy debate to the nuclear ethical debate examined in chapters 2–4 and to the diverse ethical principles that the competing sides invoked. It then turned to an alternative conditions-focused disarmament model based on the common security imperative defended in chapter 3. It emphasized the importance of vigilant antinuclear civil society activism, the acknowledgment of the prospects of deterrence failure and its catastrophic effects by the NPT member states which bear the greatest responsibility over Article VI requirements, the cultivation of security dilemma sensibility among the NWS and umbrella NNWS, and their acceptance of the common security conception as normative for international politics going forward. Finally, the chapter addressed some of the key moral challenges and dilemmas for a common security approach to nuclear disarmament: including the cultivation of nuclear fears, the exercise of security dilemma sensibility in the absence of such fears, and the choice by any N5 to send the "costly signal" of implementing key disarmament proposals in the face of uncertainties that their nuclear-armed adversaries would reciprocate.

From these separate strands of discussion, the following sketch framework of nuclear ethical maxims can be distilled: that is, all things considered, a morally responsible approach to nuclear disarmament is comprised by the following:

1. Actions that are consistent with the goal of a complete, irreversible, and verifiable nuclear abolition, to which all NPT states-parties have committed more than once;
2. Actions consistent with the principle of universalizability and which regard self and others as ends and not mere means to ends;
3. Actions that are consistent with the Principle of the Morality of Social Institutions (PMSI) or, if necessary, this principle's amended formulation; and
4. Actions whose justifications could not be reasonably rejected by N5 and NNWS alike.

The first maxim emphasizes the relationship of *consistency* between the NPT regime's goal of a complete, irreversible, and verified nuclear disarmament[120] with any member-state activity undertaken in the name of nuclear abolition. Thus, only actions consistent with this disarmament conception are morally responsible. One implication of this first point is that it is not (at least *prima facie*) morally responsible to justify the retention of nuclear deterrence in the name of facilitating nuclear abolition, as the Obama and Trump administrations have done.[121] Put simply, keeping nuclear weapons is not consistent with eliminating them. If retaining nuclear deterrence is somehow morally responsible, despite the HINW, then the United States and other N5 need to make a convincing argument that nuclear abolition is neither a morally required outcome nor an outcome to which they have voluntarily committed. Alternately, they must convincingly trace the process between retaining nuclear deterrence today and a future nuclear-weapon-free world consistent with the amended PMSI as applied to the challenge of cultivating nuclear fears. Additionally, they must trace the process from deterrence to abolition that does not ignore the HINW, the counter-fears of N5 and NNWS alike, nor which continually places the burden of initiating costly signaling solely on one's adversaries and never on oneself. The N5 have not yet traced the disarmament process in these ways, and it is reasonable to doubt that they can.

The second maxim emphasizes the deontological aspects of a morally responsible nuclear abolition. The universalizability condition is consistent with the aspirations of the NPT and TPNW, namely that each and every state has a right to be free from (the threat of) nuclear harm. This right entails a corresponding duty upon each and every state to undertake actions consistent with the goal of nuclear abolition. And, unlike the practice of nuclear deterrence *qua* nuclear hostage- holding for the sake of a given N5 state's national security, morally responsible actions toward nuclear abolition cannot hold hostage any individual or group given the common security imperative. This is to say, it is inconceivable that hostage- holding could ever be a feature of an authentic common security arrangement between or among states and peoples.

The framework's third maxim emphasizes the consequentialist aspect of morally responsible nuclear abolition. It reiterates the importance of the amended PMSI, especially in relationship to actions that appear inconsistent with its second or first maxims. Thus, the cultivation of nuclear fears arguably violates the individual human right of personal security; even so, the amended PMSI permits such action if it is indispensable to the overwhelmingly important social benefit to be derived (i.e., humanity's survival). Similarly, the expectation of reciprocation by states in the making of concessions might be inconsistent with certain understandings of national security rights. Nonetheless, any reciprocal practice of costly signaling that leads to mutual vulnerability can and has led to increased and common security for states. On this third maxim, we can infer that such action is morally responsible.

The final maxim emphasizes the common security aspect of morally responsible nuclear abolition. The requirement is that disarmament-related actions must be justifiable to both N5 and NNWS according to the principle of what each could not reasonably reject.[122] The reasonability feature of this requirement is partly a joint function of the consistency, deontological, and consequentialist elements from the first three maxims. Thus, it is *prima facie* unreasonable for any N5 to oppose nuclear disarmament measures that the first three maxims prescribe. Keenly felt or ideologically motivated objections by some of the NPT states-parties to implementing disarmament proposals or to reasonable delays in their implementation would not be morally permissible unless they satisfied widely known and commonly accepted standards of cogency or validity. The virtue of this final point is that the proposed framework for a morally responsible nuclear abolition is not required to anticipate every future contingency that might bear upon the evolution of nuclear abolition processes. The sketch framework must itself meet the consequentialist standard of utility, and therefore it ought to not strive for perfection in the anticipation of every possible scenario states might encounter.

For those who might believe that this sketch framework is unacceptably biased toward the nuclear abolitionist position, it is important to remember the political and legal obligations expressed by NPT Article VI and the 1995, 2000, and 2010 RevCon Final Reports to which each N5 and NNWS have voluntarily joined. The NPT cannot indefinitely tolerate nuclear deterrence and remain a *legitimate* security regime. The refusal of the N5 to endorse language consistent with the HINW during and after the 2015 NPT RevCon Final Report cannot erase their previous acknowledgments of the humanitarian effects of nuclear war or accident. Taken together, the NPT disarmament commitments and the HINW sufficiently establish the morality of the nuclear abolition cause. The four-point sketch framework above recommends a morally responsible path toward the realization of that cause.

NOTES

1. Portions of this chapter constitute an adaptation and update of Doyle II (2015c).
2. For a discussion of enemy images, see Wheeler (2018, 75–99).
3. See Ellsberg (2017).
4. Treaty on the Non-Proliferation of Nuclear Weapons (NPT) (1968).
5. Austria (2018); Chinese Delegation (2017); Egypt (2018); Ireland (2018); Dehghani (2017); Ulaynov (2017); Wood (2017). For a complete list of states-parties' statements for the 2017 and 2018 PrepComs, see the following links, respectively: (1) https://www.un.org/disarmament/wmd/nuclear/npt2020/prepcom2017/; (2) https://www.un.org/disarmament/wmd/nuclear/npt2020/prepcom2018/.
6. Konoe and Maurer (2014).
7. Austria (2018, para. 5). See also Sauer and Reveraert (2018, 6–8).
8. Council of Delegates of the International Red Cross and Red Crescent Movement (2011).
9. On the long-standing concern of the global antinuclear movement, see, for example, Schell (1982, 1984); Shue (2004).
10. Tutu (2014).
11. For an analysis of necessity in military and moral terms, see, for example, Walzer (2015 (1977), chapter 16).
12. Tutu (2014, 4).
13. Tutu (2014, 5).
14. For a general account on the necessary and sufficient conditions of international prohibition regimes, see Nadelmann (1990).
15. Wittner (2009, 223–25).
16. New Zealand on behalf of the New Agenda Coalition (Brazil, Egypt, Ireland, Mexico, New Zealand, and South Africa) (2018). See also Walker (2012, 117–23).
17. MacFhionnbhairr (2004).

18. 2000 Review Conference of the Parties to the Treaty on the Non-Proliferation of Nuclear Weapons (2000, 14–15).

19. Jones and Marsh (2014).

20. The White House (2002). For my more detailed treatment of this iteration of disarmament politics, see Doyle II (2009); Doyle II (2015b, especially chapter 3); and Doyle II (2017a).

21. Johnson (2005).

22. See, especially, Obama (2009).

23. 2010 Review Conference of the Parties to the Treaty on the Non-Proliferation of Nuclear Weapons (2010). See also https://dfat.gov.au/international-relations/security/non-proliferation-disarmament-arms-control/policies-agreements-treaties/treaty-on-the-non-proliferation-of-nuclear-weapons/Pages/2010-npt-review-conference-64-point-action-plan.aspx. For the twenty-two steps related specifically to nuclear disarmament, see New Zealand on behalf of the New Agenda Coalition (Brazil, Egypt, Ireland, Mexico, New Zealand, and South Africa) (2018).

24. Mukhatzhanova (2014, 5).

25. Austria (2018, para. 5).

26. The list of Annex 2 states is found on an archived web document at the United States Department of State, https://2009-2017.state.gov/t/avc/rls/159264.htm, accessed on 27 December 2018.

27. Doyle II (2017a, 16–17).

28. Austria (2018, para. 6).

29. The New Agenda Coalition (2017).

30. The New Agenda Coalition (2017).

31. The New Agenda Coalition (2017).

32. Group of Non-Aligned States Parties to the Treaty on the Non-Proliferation of Nuclear Weapons (2017).

33. The official document registering the votes of the participating states for the TPNW vote in the UNGA is located at https://s3.amazonaws.com/unoda-web/wp-content/uploads/2017/07/A.Conf_.229.2017.L.3.Rev_.1.pdf. Accessed on April 1, 2018. See also Wright (2015).

34. Lennane (2014).

35. Algeria et al. (2018).

36. The International Campaign Against Nuclear Weapons (ICAN) keeps an updated record of state signatories and states-parties to the TPNW. It is accessible at http://www.icanw.org/status-of-the-treaty-on-the-prohibition-of-nuclear-weapons/.

37. Austria (2018); Ireland (2018); New Zealand on behalf of the New Agenda Coalition (Brazil, Egypt, Ireland, Mexico, New Zealand, and South Africa) (2018).

38. See Sauer and Reveraert (2018).

39. See, for example, Ford (2018); Haspels (2017).

40. Chinese Delegation (2017); Ulaynov (2017); Haspels (2017).

41. See, for example, Wheeler (2018).

42. Chinese Delegation (2017); China (2018).

43. Chinese Delegation (2017). It is therefore interesting to note that, while Beijing pressed its case against U.S. nuclear policy, they were also assisting North

Korea's intercontinental ballistic missile program designed to deter Washington from regime change operations against Pyongyang. See Woodward (2018, 178–82).

44. Ulaynov (2017).
45. Ulaynov (2017).
46. United States of America (2018).
47. United States of America (2018).
48. Wood (2017).
49. For an extended argument on this point, see Roberts (2016). See chapter 3 for a more complete discussion of Roberts' book.
50. China (2018); Ulaynov (2017); Wood (2017).
51. Ford (2018, 9). Emphasis is in the original.
52. Haspels (2017).
53. Nuclear Threat Initiative (NTI) (2018).
54. Non-Proliferation and Disarmament Initiative (NPDI) (2018, para. 7).
55. Non-Proliferation and Disarmament Initiative (NPDI) (2018, para. 9e–f).
56. Non-Proliferation and Disarmament Initiative (NPDI) (2018, para. 10).
57. European Union (2018, paras. 12–20).
58. European Union (2018, para. 13).
59. European Union (2018, para. 15).
60. European Union (2018, 20).
61. Ford (2018).
62. Ford (2018, 5).
63. Ford (2018, 9).
64. Wittner (2009, 223).
65. See also Walker (2012, 5–6).
66. Wittner (2009, 224).
67. Nadelmann (1990).
68. Nadelmann (1990, 484–85).
69. Walker (2012, 53–85).
70. Nadelmann (1990, 493–94).
71. For a recent treatment on the potential of the TPNW to contribute to stigmatization of nuclear weapons and NWS, see Sauer and Reveraert (2018). For the nuclear use taboo, see Tannenwald (2007).
72. Nadelmann (1990, 484–85).
73. Walker (2012, chapters 2–4).
74. Dunn (2009); Fuhrmann and Lupu (2016).
75. Tannenwald (2007, 2018).
76. This was the near-consensus opinion of several scholars and policy experts at the "Response to North Korean Nuclear First-Use Workshop" convened by James Scouras and Erin Hahn at the Johns Hopkins University's Applied Physics Laboratory in Baltimore, MD, April 22–24, 2019. This author was one of the invited participants.
77. Nye (1986, 61–62).
78. Gavin (2012, 147).
79. Gavin (2012, 39–41).
80. See, for example, Woodward (2018, 305–8).

81. Booth and Wheeler (2008, 167–68); Wheeler (2018, 76–82).
82. Booth and Wheeler (2008, 4–6).
83. China (2018).
84. Walzer (2015 (1977), 250–54).
85. Booth and Wheeler (2008, 150).
86. Booth and Wheeler (2008, 153); Walker (2012, 99–102).
87. Booth and Wheeler (2008, 155).
88. Deutsch et al. (1957).
89. Palme (1982a, emphasis added). Also quoted in Booth and Wheeler (2008, 138–40).
90. Wittner (2009, 182–83).
91. Booth and Wheeler (2008, 146–48).
92. Doyle II (2015c).
93. Wittner (2009, especially chapters 5 and 9).
94. Booth and Wheeler (2008, 229).
95. Sauer and Reveraert (2018, 14).
96. International Campaign to Abolish Nuclear Weapons (2017).
97. See, for example, Algeria et al. (2018) Austria (2018); Council of Delegates of the International Red Cross and Red Crescent Movement (2011); Group of Non-Aligned States Parties to the Treaty on the Non-Proliferation of Nuclear Weapons (2017); Ireland (2018); Konoe and Maurer (2014); New Zealand on behalf of the New Agenda Coalition (Brazil, Egypt, Ireland, Mexico, New Zealand, and South Africa) (2018); Tutu (2014).
98. Walzer (2015 (1977), 268).
99. Quinlan (2009, 50–51).
100. Schell (1982, 1984). See also Lee (1985, 555).
101. For the history of nuclear testing, see Comprehensive Nuclear-Test-Ban Treaty Organization Preparatory Commission (2018).
102. Booth and Wheeler (2008, 149); Hoffman (2009, 90–96).
103. Kant (1996, 78–80).
104. Nye (1986, 99).
105. Walzer (2015 (1977), 273).
106. Lee (1985, 551).
107. Lee (1985, 558).
108. For a detailed account, see Wheeler (2018, 143–91).
109. Davenport (2018).
110. See, for example, Booth and Wheeler (2008, 65–70).
111. Lorber (2016).
112. Reif (2018); Landler (2018).
113. See Ford (2018).
114. Booth and Wheeler (2008, 91); Kydd (2000).
115. Booth and Wheeler (2008, 169).
116. Hoffman (2009, 265–66).
117. This is, after all, the same kind of argument made by nuclear deterrence advocates: i.e., that nuclear deterrence is morally justified (or morally responsible) given its success at preventing superpower nuclear confrontation during the Cold War which to many seemed inevitable.

118. See, for example, Wheeler (2018, 69–72).
119. See, for example, Wheeler (2018, 69–72).
120. Sauer and Reveraert (2018, 8).
121. U.S. Department of Defense (2010); Office of the Secretary of Defense (2018).
122. Scanlon (1998, 103–6, 191–97).

Chapter 6

Conclusion

A Nuclear Ethics of Common Security Orders

The preceding chapters explored issues and advanced arguments in support of a core contention originally advanced in chapter 1: that a twenty-first-century nuclear ethics must prioritize the considerations of fundamental or rudimentary justice for an international nuclear order whose core purpose is to ensure the survival and security of all states and peoples. This contention has been formulated as Justice > Order > Survival. This formula might be interpreted causally: that is, the survival of states and peoples is the condition produced by the construction of a just international nuclear order. It might also be understood as a statement of the properly ranked order of values independent of their causal relationships. This is to say, justice is a concept that summarizes cherished notions of interpersonal and interstate relations organized around fair distributions and dealings, equality of opportunities, and reciprocity among equals in various types of political society. Thus, justice is a duty each actor owes to every other actor such that the relevant social order achieves its purpose of ensuring the survival and security of its members. In the context of the current international nuclear order as shaped by international political and legal dynamics, this means that decisions on nuclear weapons policy among nuclear-weapon states (NWS) and non-nuclear-weapon states (NNWS) must be ordered around the considerations of fundamental justice.

The focus on fundamental justice among states in this book's core argument is partly a response to nuclear ethical approaches that privilege other values, such as the maintenance of an international nuclear order designed around the interests of the world's most powerful states. Admittedly, Kantian-inspired commentators privileged principles of justice among states and peoples in their absolute condemnations of superpower nuclear war plans and deterrence policy during and after the Cold War. As chapter 2 related, their condemnations rested on the justice owed to individual persons regarding their fundamental rights of life and liberty. For instance, Thomas Donaldson argued that nuclear warfighting ignored the proper limits of national self-defense by disregarding these fundamental individual rights, and Steven P. Lee contended that nuclear deterrence had become an unjust policy of hostage-holding.[1] And as chapters 3 and 5 related, after the Cold War's end, similarly inspired commentators advanced the Humanitarian Imperative to Abolish Nuclear Weapons (HINW or the "Humanitarian Imperative"), an ethical principle expressing the justice claims of humanity's rights of survival and security against the narrow national security interests of the nuclear-armed states and, of the specially important, NPT NWS (also known as the N5).[2] Clearly, if every human being deserves to live and enjoy security, then it would be incorrect to say that the Kantian-inspired commentators have overlooked the priority of justice in nuclear ethics.

One might even argue that some commentators defending the morality of nuclear defense and deterrence have prioritized the value of justice over mere order. After all, national security is accepted as a just cause (*jus ad bellum*) inasmuch as each state deserves to survive as a sovereign political community. It follows that aggression is an injustice or "crime" that must be defeated, and this must be typically undertaken by force of arms.[3] Additionally, many commentators invoking just war reasoning believe that extreme measures of national self-defense are permitted or required in conditions of existential threat,[4] including the use or the threat of the use of nuclear weapons.[5]

At some point, however, one might reasonably conclude that no qualitative difference exists between a nuclear-armed international society with legal *jus ad bellum* provisions and a Hobbesian anarchic system of states where international legal structures are entirely absent and fundamental justice itself is practically impossible to realize.[6] In both worlds, conditions of existential threat entail that one state's survival might well come at the expense of another's. And while it might appear that international justice is possible among states that have chosen international law as a medium for their interactions, in conditions of existential threat these structures, and any conception of justice among states that attend them, dissolve in practice and leave states to the mercy of the rule of force. For such worlds, therefore, the proper expression would be Order > Survival.

Some expressions of nuclear ethics affirm this Hobbesian formula to greater or lesser degrees. Chapter 2 related approaches that identified morality with rationality, which defended the permissibility of nuclear deterrence to dissuade nuclear aggression, and which affirmed the rationality of nuclear reprisal to restore deterrence after nuclear first-use.[7] Indeed, chapter 4 argues that even liberal defenses of nuclear deterrence cannot ultimately avoid succumbing to security measures that are illiberal and resistant to the rule of international law.[8] Accordingly, we might infer from chapter 5's discussion that the U.S.-motivated Conditions-Focused Disarmament Discourse[9] aligns with a "liberal" Hobbesianism. Accordingly, a suitably modified formula for Rawls's liberal nuclearism or the U.S.-motivated Conditions-Focused Disarmament Discourse might be expressed as Liberal Order > Liberal Survival.

In short, seven decades of nuclear ethical debate have not produced a consensus position on the morality of nuclear war, deterrence, arms control, or disarmament. While enduring moral dissensus is neither surprising nor unusual in domestic and international politics, it is not thereby acceptable nor necessarily free from justifiable criticism. As long as moral consensus evades us, concerted political action is unable to proceed with moral clarity. Thus, is there a consensus in nuclear ethics to be had? Is there a way to reconcile or harmonize approaches, the principles and maxims of which appear irreconcilable? The previous chapters have responded affirmatively to this question by asserting that the most fruitful nuclear ethics for the twenty-first century would be anchored on the common security conception understood as a moral imperative. The introduction of the common security conception is meant to identify the kind of interstate dynamic that would comport with a just international order. This chapter's task is to explicate that nuclear ethics of common security in more detail. Before this task can be directly undertaken, however, we must first understand better the underlying dilemma for nuclear ethics, which is what I call the "order-justice dilemma."

THE ORDER-JUSTICE DILEMMA

William Walker's 2012 account of the history of the international nuclear order is prefaced in part by the observation that "human survival has rested on a gamble that relations between nuclear-armed states, their behavior towards other states, and their handling of nuclear assets can be well enough managed to avoid—perpetually—the outbreak of war by accident or design."[10] In this statement, we find an explicit identification of the intrinsic moral value of human survival and the fundamental instrumental value of political order—that is, instrumental insofar as order is conceived as necessary for human survival. Later, Walker writes,

> My starting point is that, beyond basic survival, the achievement of *order* is—and has to be—the pre-eminent and perennial concern of states, and especially of the great powers among them, given the existence of this ultimate instrument of destruction and symbol of state power.[11]

The concept of "management" in Walker's former statement implies the concept of "order" in his latter statement, and the question of orderly relations among NWS is cast as crucial to the prevention of conventional wars that might lead to nuclear war that, in turn, could lead to the extinction of humanity. However, some ways of managing nuclear assets, or of managing interstate relations generally, will advantage some states at the expense of others and without regard to fundamental legal or moral notions of fairness and reciprocity. In other words, there is a distinction between just and unjust international orders, and the question arises as to whether human survival depends upon international order *per se* or rather upon a just international order.

The order-justice dilemma is the collective predicament of NWS and NNWS where officials representing states and international organizations are collectively uncertain whether international order *per se* ought to be prioritized over international order framed around principles of justice. Chapters 2–5 showed that the NWS and their allies have largely grasped the former horn of the dilemma while the abolitionist NNWS and civil society groups have grasped the latter horn. This lack of consensus does not mean that international society has not defaulted to one of these two orientations. Arguably, the N5's power and influence have determined that nuclear restraint defines the international nuclear order, leading to the marginalization or exclusion of NNWS' justice claims. This book's argument is that privileging the order of nuclear restraint at the expense of fundamental justice claims on nuclear matters is deeply problematic; indeed, the tolerance of nuclear injustices risks the ultimate objective of the nuclear order: the survival of humanity.

One might object that an absolute insistence on the priority of fundamental justice among states will destabilize relations and perversely lead to grave insecurities or harm. Critics of the Treaty on the Prohibition of Nuclear Weapons (TPNW) might, as chapter 5 suggested, accept that nuclear abolition is an important goal, but that strict adherence to legal and political disarmament commitments without due regard for N5's strategic and security dilemmas can invite nuclear cheating and cause the very nuclear conflict that nuclear abolitionists want to avoid. On this view, strategic stability and the corresponding order of nuclear restraint must first be established before any movement toward the justice of honoring treaty obligations can be realized. Let us now carefully consider two general theoretical accounts that seem to support this objection.

The first account is from Russell Hardin, whose central contention was that the "achievement of general social order comes prior to justice, democracy, and other systemic achievements."[12] On Hardin's view, conceptions of justice vary among individuals and communities within modern domestic and international society, and this means that moral dissensus is normal and not necessarily a condition to avoid or change. Consequently, for any person or community to attempt to impose their justice conception on society at large, especially if they felt aggrieved, would be destabilizing and constitute a threat to domestic order and security. Thus, the social priority must be order among diverse communities with competing conceptions of justice. For Hardin, it is only in the confined context of an ancient or medieval village where rudimentary and minatory principles of justice can take priority and shape the social order. Principles such as "do not break promises," "do not lie," and "do not steal" were central in village life and were reinforced by its conditions of social intimacy.[13] Hardin imagines a villager named Bodo whose life is governed by such minatory principles, and he calls this normative system "Bodo Ethics." The internalization of these rudimentary norms as binding on villager conduct did not require elaborate conventions such as codified law. By contrast, Hardin argues that Bodo Ethics are not able to anchor "a principle of quasi-egalitarian distributive justice in a large and complex society" where social intimacy can no longer function as an effective regulatory mechanism.[14] On his view, large and complex societies might turn to Bodo Ethics as a starting point for social cooperation; however, these rudimentary norms must be adapted significantly to fit the complexities that attend domestic political and social relations within or among large modern states.

Hardin elaborates on the preceding point by identifying social contractarianism as a significant adaptation of Bodo Ethics to modern society. Early modern European contractarians from Hobbes to Kant imagined individuals dwelling in more or less dangerous anarchic environments (i.e., the state of nature) in which their fears and insecurities motivated the formation of the state wherein their personal security was obtained in exchange for their (near) absolute subjection to the constituted sovereign.[15] This moment of state formation thus becomes the basis of political society where Bodo's norms are applied and refined by positive law and a judicial system no longer tethered to the intimate accountability mechanisms of village life. Moreover, once states adopt legal instruments to regulate interstate security or economic relations,[16] it becomes possible to conceive each state's moral obligation to honor its treaty or legal commitments—that is, *pacta sunt servanda*.[17] Hence, the constitution of an international legal order establishes the possibility of international justice understood as the satisfaction of states' legal rights and obligations in relation to other states. Within this broad international legal order, the set of bilateral and multilateral treaties that comprise the nuclear arms control and disarmament regime establishes the "international nuclear order."[18] And, as we have seen, the key questions at the heart of the international nuclear order involve not just questions of national interest but questions of duties to honor nuclear arms control and disarmament treaty obligations. These, too, are questions of international justice, but Hardin's view suggests that the establishment of the international nuclear order is primary while considerations of international justice within that order are secondary.

That questions of international justice have arisen in the international order explains the concern regarding unjust orders and their corresponding dangers.

Although members of domestic and international political society retain self-interest as a central motivation for action, one might reasonably assume that social contractarian principles, such as political obligation, can function as effective sources of motivation. Accordingly, we might expect political actors to be intrinsically motivated to honor their contractual or legal obligations. Unfortunately, Hardin finds the opposite is true.[19] He argues that the history of social contractarian societies shows that many members often choose free riding strategies (e.g., tax cheaters). And in many other cases, it is self-interest that drives members' decisions to honor their obligations. The implication of Hardin's analysis is that the consent that self-interest gives within social contractarian societies is the consent that self-interest can also take away.

The centrality of self-interest as a persistent and overriding motivation for members of social contractarian arrangements thus explains why promise-keeping is often treated as a conditional or nonbinding obligation.[20] As Hardin puts it,

> In essence, when we design institutions of justice, we include within them devices to give strong incentives for compliance that, without the incentives, would fail. Such incentives are *de facto* naturally built into promise-keeping in our world and in virtually all interactions in [the] transparent world [of a small village], but they commonly must be added onto the incentives for contributing to a collective action.[21]

Hardin's remarks emphasize the role of self-interest in village interactions as well as complex modern social interactions. The medieval villager and modern political actor alike are interested in honoring obligations to secure material and reputational gains and to avoid any material or reputational costs that might lead to their exclusion from (the benefits of) future social interactions. In modern society, the mode of the contract substitutes for the mode of village intimacy as the effective means of channeling self-interest and holding members accountable.

The upshot of Hardin's remarks is that, in village and modern society alike, the establishment of social order must be the priority. Once social order is achieved and there is significant assurance that people holding diverse views on justice will nonetheless cooperate to achieve important social goods, then the moral salience of intention or motivation diminishes in favor of the moral salience of achieving good outcomes. In short, "I may be a Kantian moralist about promise-keeping while you may be a Humean pragmatist. . . . We will [nevertheless] succeed."[22] If Hardin's account is correct, then it explains the enduring social force of moral consequentialism in international ethics and, more narrowly, nuclear ethics. Moreover, it seems to strengthen the claim in favor of the formula Order > Survival.

A second account that proposes the priority of social order is Terry Nardin's account of Kantian republicanism.[23] According to Nardin, Immanuel Kant's moral and political theory is akin to Hobbesian theory insofar as it is unable to ground an argument that international justice is possible in an anarchic world of security-seeking states. Indeed, Nardin claims that "politics for Kant is not applied ethics. Its subject matter is different, and its principles are independently grounded."[24] Specifically, Nardin contends that Kantian ethics is grounded on the categorical imperative as the formal expression of the supreme law of moral reason, while Kantian politics is grounded on the doctrine of right in which each citizen of a constitutional republic has the right of freedom from any exercise of arbitrary power or domination by others (freedom as independence). For Nardin, Kant's political theory, then, emphasizes justice as a political rather than a moral concept.

Accordingly, for Nardin a theory of justice for Kant "identifies principles of civil or international order, not the institutions needed to secure these principles in contingent circumstances,"[25] and, necessarily, a theory of justice is a "theory of the justice of the law, and where no law exists there is nothing to discuss."[26] In conjunction with the doctrine of right, Nardin contended that Kant's conception of justice is defined as the moment when the state may appropriately exercise coercive power against any individual or group within its jurisdiction that wrongfully interferes with any citizen's right of independence.[27] Clearly, to define justice as the justice of laws against wrongful incursions of individuals' freedom rights is to presuppose an established civil order, but it is also to presuppose the deontological foundation of that civil order. Accordingly, in some sense we might read Nardin's take on Kant as allowing for the equal importance of order and justice for civil society.

Regarding international society, Nardin recalls Kant's contentions in *Perpetual Peace* that one of the necessary conditions of a pacific international order is that each participating state have a republican constitution and that they settle their disputes on the basis of law.[28] On Nardin's view, the unfortunate implication of these contentions is that the term "justice" is not applicable even for a Kantian pacific international order. As Nardin states,

> Justice is possible only if the members of a society, either civil or international, can coexist under laws that respect their independence and secure it by protecting them from internal domination or foreign aggression. Peace-with-rights is a good that is internal to a system of justice. It is a peace based on rights defended by recourse to law, not force. But, strikingly, the pacific federation Kant envisions is a voluntary one that allows for secession; it is not, like a state, a nonvoluntary legal order, an association on the basis of coercive laws. It may provide peace, but because that peace is not legally secured, it is not a peace-with-rights.[29]

For Nardin's Kant, the term "peace-with-rights" describes the condition under which each citizen-subject of a constitutional republic or state is no longer at war with any other citizen by virtue of the binding civil law to which each citizen is voluntarily subject. Such citizens are not free to withdraw from their civil obligations and still rightfully claim protection from domination by the civil authorities. However, states-parties to a Kantian pacific federation are not "citizen-subjects" of that international order. They remain sovereign actors and may rightfully secede from that federation at any time. Any peace that endures among these states-parties is not by virtue of international law and its corresponding rights, but rather by virtue of shared identity, culture, or interests.

Accordingly, Nardin interprets Kantian international theory and IR realist theory as more compatible than many Realists grant. Specifically, both agree on the nature of the anarchy condition and its effect on international law. Nardin states,

> In the absence of a superior authority to fix the meaning of legal rules and enforce them, the meaning of its rules as they are applied in particular situations, and therefore, the rights and duties they prescribe cannot be authoritatively determined.[30]

For Nardin, the Kantian pacific federation is not a world government or state of states such that each state-party is subject to a central governing authority that can end controversies over international legal language and thus fix and enforce states-parties' obligations. Instead, any controversy over the meanings of treaty, charter, or convention language remains formally indeterminate and subject to interpretation and even subversive applications by sovereign states. In this regard, Nardin's account of Kantian anarchy seems entirely consistent with mainstream IR realist accounts.

Nardin's Kant would therefore not be surprised to observe the never-ending disputes among the N5 and the abolitionist NNWS concerning the meanings of the terms in NPT Article VI on nuclear disarmament. Indeed, Nardin's Kant might argue that the competing understandings of Article VI language is a function of conflicts of national interest that, as Russell Hardin argued above, constitutes the foundation of justice claims that abolitionist NNWS make against the N5 and that the N5 reject as invalid. From this inability of states to fix the terms of international law generally and the NPT in particular, we find that the "obligations prescribed by the law of nations . . . are, Kant says, provisional—they are moral obligations awaiting incorporation into a positive legal order."[31] It follows on this view that the "problem is not that perpetual peace [or nuclear abolition] between states—a full, secure, and non-voluntary system of justice that, nevertheless, preserves their sovereignty—is contingently unlikely but rather that it is conceptually impossible."[32]

If Nardin's interpretation of Kant's international theory is correct, then it would join with Hardin's account in affirming the formula Order > Survival. It is therefore quite interesting that recent empirical analysis constitutes a challenge to their conclusions and, instead, affirms this book's contention that Justice > Order > Survival. In their study of sixteen civil wars during the 1990s, Daniel Druckman and Cecilia Albin found that a durable peace agreement between armed domestic factions is much more likely if principles of distributive justice framed the reconstitution of the state's political and social order that had been previously destroyed.[33] Druckman and Albin define "durable peace agreement" as a function of the fidelity of the parties to the agreement and the length of time the agreement lasts. They find that durable peace agreements can suffer some problems of noncompliance by one or more of the factions but not in a way to undermine or subvert the agreement's central objectives.[34] Most importantly, their study showed that the inclusion of distributive justice principles leads to durable peace agreements unless there are low levels of trust among the participants or unless the nature of the civil war was ranked extremely high in intensity.[35] Moreover, the most durable agreements occurred when the relevant parties linked their privileged principle of equality to their need to change the political order or to establish "new collaborations as a means of new relationships in the future."[36]

One might wonder if Druckman and Albin's findings are generalizable to the questions of international politics and nuclear order. One positive indication is suggested by their citation of the 1987 Montreal Protocol on Substances that Deplete the Ozone Layer as an exemplar of a durable international agreement that privileged principles of international justice.[37] They identify three such principles. One is the *proportionality principle*, which led states- parties to roll back chlorofluorocarbon (CFC) emissions back to 1986 levels. The collective commitment to CFC reductions imposed greater costs on the industrialized states than what developing states would have to pay, since the former's emissions had contributed the most to the depletion of the ozone layer. And yet, these states-parties accepted greater costs in the name of the collective good. A second principle was the *compensation principle*, where financial assistance was awarded to global South countries so that they could comply with their commitments. It also authorized the collective decision to allow global South countries a ten-year exemption on compliance so that they might catch up with the developed global North states. Finally, the *equality principle* stated that once parity was reached in the CFC emissions of each state-party, then each would share the costs of treaty compliance equally with all the others. Had not these principles been woven into the heart of the Montreal Protocol, Druckman and Albin contend that it is highly unlikely the protocol would have endured and the damage to the ozone layer might well have been irreparable. And while the Montreal Protocol is not an arms control and disarmament treaty, it seems reasonable to conclude that the objective of preserving humanity from grave environmental harm is akin to the objective of preserving states and peoples from grave nuclear harm.

At the very minimum, the Druckman and Albin account, if generalizable, suggests that justice and order are equally important values to anchor a twenty-first-century nuclear ethics. At the maximum, they suggest that durable social order is dependent upon fundamental principles of justice. The question is now raised: Do the Druckman and Albin findings offer a satisfactory response to the arguments of Hardin and Nardin? Is there still good reason to argue for the priority of justice over order as such? I believe so.

First, Druckman and Albin show that the disorder or anarchy that accompanies civil war does not count as a thorough elimination of social order, even if a country's government collapses and its economy tailspins. Participants in civil war retain their socialized expectations regarding Bodo norms and the corresponding framework of social order. This is to say, the disorder of civil conflicts does not erase rudimentary ideas of justice that were taught by families, schools, and religious institutions among the individuals and peoples involved. It can be imagined that leaders and key members of armed factions, as well as relevant civil society groups, advance justice claims during peace negotiations. Moreover, it is expected that they regard the satisfaction of justice claims as a necessary condition of a future peaceful order. In the end, Druckman and Albin show that there is a statistically significant correlation between inclusion of distributive justice principles—especially the equality principle—and the durability of peace agreements.[38] Taken together, their findings imply that disorder is repaired not by order *per se* but by the emphasis on just ordering of political relationships informed by principles of equality, compensation, and proportionality.

One implication of Druckman and Albin's findings is that any theoretical analysis on the question of the priority of justice or order for questions of international security, and nuclear defense and deterrence, must be careful to reason from warranted assumptions concerning the (initial) conditions of social interaction among adversaries. The context surrounding the international politics of nuclear defense and deterrence policy, including the politics of nuclear arms control and disarmament, does not lend itself to naïve Hobbesian assumptions concerning a hypothetical state of nature and the ignorance or amorality of political actors prior to interaction. Indeed, one criticism of Hobbes's account of the state of nature is that his hypothetical actors in that natural state were imagined as seventeenth-century Englishmen possessing a rich knowledge of social order, politics, and morality.[39] Rather, one might argue that the contemporary international political and nuclear context lends itself more closely to Rousseau's assumptions concerning the negative effects of competing self-interests in a society marked by radical inequalities of power, wealth, and status that the social contract aims to resolve.[40] Indeed, it is Rousseau's social contract that aims to "bring together what right permits and what interest prescribes, so that justice and utility do not find themselves at odds with one another."[41]

In this vein, it is interesting that, in the middle of his account, Hardin proposes that "more complex and encompassing orders build on simple systems such as Bodo's. Primitive ethics can help us achieve enough order for us to build far beyond such systems."[42] For Hardin to say in the middle of his account that Bodo Ethics is a groundwork upon which complex and encompassing social orders can be constructed is to say that "primitive" conceptions of justice count as constitutive and causal preconditions of the durability of such orders. In conjunction with Druckman and Albin's analysis, it now becomes tempting to read Hardin's account as proposing that Bodo Ethics of Justice > General Social Order > Specific Applications of Justice Conceptions > Security/Survival. If this interpretation of Hardin is plausible, then we have additional reason for thinking that the book's formula of Justice > Order > Survival is also plausible.

Much more might be said concerning the Order-Justice dilemma. However, I shall assume the foregoing has adequately laid the groundwork for the next section's more detailed account of a twenty-first-century nuclear ethics of justice, order, and survival. Let us call this a "nuclear ethics of common security orders." Accordingly, the next several paragraphs draw on the arguments from chapters 2–5 to discuss more fully the basis for such a nuclear ethics and afterward relate and defend a corresponding set of nuclear ethical maxims.

A NUCLEAR ETHICS OF JUST COMMON SECURITY ORDERS

Initial Considerations

A nuclear ethics of common security orders that privileges rudimentary principles of justice should provide maxims to lay the groundwork for bilateral and multilateral security arrangements defined by mutuality (i.e., "security with") and fairness and thus intensify the momentum toward an effective and complete nuclear abolition. The groundwork for this kind of nuclear ethics must, as I see it, attend to two sets of initial considerations of justice that are applications or extensions of Bodo Ethics.

One set of initial considerations is informed by Kant's discussion of the preliminary articles for perpetual peace in the geopolitical context of Earth's finite landmasses. Kant argues that all human beings have a right in common to the Earth's surface and to its resources from which their lives depend, which means that no one has an overriding human right over another person to claim or use any specific piece of land (or water) to survive. Moreover, the finiteness of habitable land will compel an expanding population of human beings to eventually "put up with being near one another."[43] Although conflicts of interest among states and peoples historically have been addressed through war, Kant believed that war's brutality would eventually compel states and peoples to enter into law-like relations in order to ensure survival.[44] Kant's preliminary articles for perpetual peace constitute a set of maxims to lead warring states into an understanding of how to lay the groundwork for the possibility of law-like and enduring pacific relations.[45]

In this vein, a first application of a nuclear ethics of common security orders stops short of prescribing Kant's definitive articles for perpetual peace[46] (i.e., each state has a republican constitution, each republican state forms a pacific federation, and each state is to honor the cosmopolitan right of hospitality) and instead advances three (amended) preliminary articles from Kant's account to questions of nuclear defense, deterrence, and arms control policies in anticipation of the construction of durable peace arrangements and a corresponding nuclear abolition. Those articles are as follows:

1. "No treaty of peace shall be . . . made with a secret reservation of material for a future war."[47]
2. "No state shall forcibly interfere in the constitution and government of another state."[48]
3. "No state at war with another shall allow itself such acts of hostility as would have to make mutual trust impossible during a future peace."[49]

These articles constitute a Kantian application of Bodo Ethics at the international level of analysis. Each expresses a distinct rudimentary minatory norm. The first proscribes state adversaries from acting in a fundamentally *dishonest* or *subversive* way such that secret advantages in capabilities are maintained against one another in anticipation of the initiation or resumption of nuclear arms racing or military conflict.[50] The second proscribes *aggression*, which is the central minatory norm of domestic and international society.[51] The third proscribes acts that establish a state's *untrustworthiness* in relation to its adversaries and, consequently, which work to subvert attempts to establish a durable peace. By acting dishonestly, aggressively, and distrustfully in contexts of security competition or warfare, a state actor sustains the kinds of injustices that produce for their adversary a *jus ad bellum* cause for continuing or resuming armed conflict. In contrast, their mutual adherence to these rudimentary minatory norms establishes a condition of basic justice such that provisional peace might be realized if a suspension of hostilities is followed by modest successes in arms control arrangements. In turn, these can be leveraged into disarmament arrangements if the common security imperative is mutually adopted.

A second set of initial considerations is suggested by considering John Rawls's conception of justice as fairness. Rawls defines fairness as a principle that requires each actor to contribute to the construction of a social good in the context of their social arrangements. Thus, the main idea of a principle of fairness is

> that when a number of persons engage in a mutually advantageous cooperative venture according to rules, and thus restrict their liberty in ways necessary to yield advantages for all, those who have submitted to these restrictions have a right to a similar acquiescence on the part of those who have benefited from their submission. We are not to gain from the cooperative labors of others without doing our fair share.[52]

This main idea contains two important points.[53] *First*, the principle of fairness is anchored on a principle of equality which prescribes that each member of a class of actors has a presumptive right to the same kind of status recognition as every other member and, thus, deserves to be treated the same as every other member. Applied in more formalized terms to the international nuclear order, the first point says that states possess the status property of sovereignty S within the NPT regime. This is not to say that the members of any given class of actors are not unequal in important ways. Not all states possessing S are powerful (P), wealthy (W), technologically advanced (A), or nuclear armed (NW). However, by the terms of international society as defined by the United Nations Charter, each state possessing S deserves equal status recognition and the corresponding equality of treatment by and with every other NPT member state. This is to say, justice as fairness gives preference to the principle of equality, all things considered.

An important corollary to the principle of equality is a principle of differentiated responsibility among equal members of a social arrangement, who are otherwise unequal in significant ways. This principle prescribes that for two or more actors that possess the same property—for example, two or more states possessing S and having membership in the NPT—that the same rights, duties, and treatments apply to them unless the possession of one or more other properties adequately justifies their enlargement or diminishment. For instance, only five NPT member states possessing S also possess NW while all other NPT states-parties do not. On this corollary to the principle of equality, their differences in the possession of NW generate differences in the duties that the N5 bear and those which NNWS bear, despite their equality in possessing S. Thus, if the possession of NW by some S states leads to greater dishonesty on their part, or aggression, or untrustworthiness in relation to their nuclear-armed adversaries or even the community of abolitionist NNWS, the N5 must bear an enhanced duty to the rudimentary minatory norms under discussion than do the NNWS. On the other hand, the possession of NW by some S states does not excuse them from the duties that they bear as NPT member states to adhere to their treaty obligations and promises. To be fair or just, any state's exclusion from NPT-related duties on S- or P-related grounds must be based on reasons that are acceptable to N5 and NNWS generally. Otherwise, the rightful equality of members on NPT-related criteria is diminished or dissolved. As a result, it is reasonable to argue that the N5 have a special, or differentiated and heavier, responsibility with regards to nuclear abolition, even though NPT Article VI assigns this as a general duty to all states-parties.

Second, it is important to highlight that Rawls's principle of fairness is also anchored on a principle of reciprocity. This principle is explicit in Rawls's contention cited above that "we are not to gain from the cooperative labors of others without doing our fair share." In the more formalized terms employed above, we might say that the rights or duties applied to one state on S-related matters ought to be applied in turn to every other state within a given international arrangement if such applications cannot be simultaneous. Indeed, the principle of reciprocity increases in importance regarding fairness and justice as the time-span lengthens between an actor's exercise of duty and the subsequent dutiful exercises of all other relevant actors. The unfairness of unreciprocated action by some members of a social arrangement is sufficient at some point to transform it into an unjust arrangement. Thus, chapters 2–5 noted that the NNWS increasingly found that the NPT regime had become grossly unfair and unjust precisely because their adherence to regime norms of nonproliferation and disarmament was not reciprocated by the N5 in equal measure. If P-related differences may not rightly excuse the N5 from NPT-related obligations, then the N5 owe the NNWS reciprocal adherence to the disarmament commitments of NPT Article VI as defined and clarified by the 13 Points and the 2010 Action Plan.

Maxims of a Nuclear Ethics of Common Security Orders

The discussion so far suggests four maxims of nuclear ethics as the twenty-first century unfolds:

(1) The common security of states and peoples in regional and global contexts is an unconditionally just cause.
(2) Under no circumstances should states undertake nuclear first-strikes or nuclear reprisal strikes.
(3) Each state should stigmatize nuclear weapons.
(4) Each measure taken by states to address the questions of nuclear arms control and disarmament must conform to principles of fairness.

Taken together, these four maxims provide a systematic ethical and political approach to the central nuclear-related security conundrums of the twenty-first century. As it is hinted previously, the common security imperative is a defining feature of this nuclear ethical code. And while the second, third, and fourth maxims do not depend upon the first one in causal terms, it seems right to believe that embedding common security into state interactions makes more likely the success of these other maxims.

Maxim 1: The common security of states and peoples in regional and global contexts is an unconditionally just cause.

In chapters 3–5, the common security imperative was proposed as a moral principle upon which a new or alternative nuclear ethics might be anchored. It is now time to weave that proposal into this chapter's argument on the priority of justice in conjunction with the principle of fairness.

Common security is distinguished by the claim that adversaries will find durable security "with" instead of "against" each other. Contrasting "security against" conceptions—such as national security, alliance security, and even some cooperative security conceptions—share the assumption that armed force or threat of armed force is required to be free from (the threat of) aggression. Of course, this assumption is valid insofar as enmity defines the regional or global dynamics among states. Accordingly, "security against" conceptions provide strong motivation for state actors to violate international society's rudimentary minatory norms if their vital interests are threatened. If "security against" thus continues to determine state security policy, then the minatory norms prohibiting dishonesty, aggression, and untrustworthiness will lack the required motivational force required for a common peace. Furthermore, "security against" commitments will override in practice any duty of justice as fairness in international politics. Specifically, "security against" will be taken to override duties arising from the principle of equality because the national or alliance interest will be taken to supersede the rightful claims of other sovereign states. "Security against" will also be taken to override considerations arising from the reciprocity principle because the rightful claims of other sovereign states will not receive their recognition. For these reasons, "security against" conceptions function as a principle of conflict, instability, and disorder in international society. Paradoxically, the more states seek order and stability through "security against" policies, the more regional or global stability and order is put at risk or undermined. In this nuclear age, "security against" is the seed of intractable security dilemmas and paradoxes alike.

As a contrast, the "security with" conception is nearly or entirely consistent with the elements of justice as fairness discussed above. Common security among adversaries affirms and strengthens each state's recognition of the other(s) in terms of equal rights of survival, security, and sovereign status. Moreover, common security affirms that each state so related should accept the burden of shared responsibilities and the liberties accompanying their shared rights. Indeed, it is this mutual recognition of burden sharing that makes international arms control and disarmament agreements more likely to succeed, as had been the case (until very recently) with the 1987 Intermediate Nuclear Forces agreement.[54] Additionally, the "security with" conception institutionalizes the reciprocity of cooperation among (former) adversaries much more than any "security against" conception. It is important to note that "security with" conception does not guarantee the continuation of trust relationships among former adversaries, but the chances of maintaining such relationships are much greater than they would be otherwise.[55]

If common security in fact exhibits these elements of justice as fairness, then the international order that (former) adversaries have constructed will be characterized as a just order, the durability of which should be much greater than any "security against" arrangement could produce. When applied to the questions of nuclear ethics, "security with" can synthesize the relevant consequentialist and deontological moral requirements on the questions of nuclear defense, deterrence, arms control, and disarmament. First, the realization of common security satisfies the consequentialist requirement of the avoidance of nuclear catastrophe. If each NWS or nuclear-aspirant state finds their security bound up with that of their adversary, then the mutuality of their security arrangement forestalls any resort to nuclear weapons for deterrence, defense, or coercive diplomacy. Indeed, any subsequent resort to nuclear deterrence, defense, or coercion would be clear proof of the dissolution of the common security arrangement and a return to a distrustful "security against" relationship. One implication of this line of reasoning is that it satisfies Nye's maxims to reduce risks of nuclear war in the short term and to reduce reliance on nuclear weapons over time.[56] This is due to common security partners' mutual decision to remove nuclear weapons from their relationship in favor of legal or political mechanisms of mutual benefit.

Second, common security is much more likely to satisfy Nye's deontological requirement that nuclear weapons must not be normalized.[57] This means that the de-normalization or stigmatization of nuclear weapons as "beyond the pale" can be accomplished without the corresponding fear of radical national or alliance vulnerability discussed earlier, since the source of each NWS security is not found in nuclear weapons anymore. This point will enjoy greater elaboration later when the third maxim is more fully discussed.

Last, common security is much more likely to satisfy Nye's deontological requirement that innocent people must not be harmed.[58] The mutual decision to de-link security policy from nuclear defense and deterrence, and the corresponding decision to treat nuclear weapons as "beyond the pale," leads to the *de facto* elimination of nuclear hostage-holding and the possibility of indiscriminate nuclear warfare. Once nuclear weapons become irrelevant by every practical measure, then the step toward universal nuclear abolition will not provoke the fears of human insecurity that the survivors of the Hiroshima and Nagasaki atomic attacks, as well as the survivors of atomic testing in indigenous territories, have suffered.[59]

Maxim 2: Under no circumstances should states undertake nuclear first-strikes or nuclear reprisal strikes.

While the rationalist or just war theoretic justification of preemptive or retaliatory nuclear use is grounded on the exclusive right of national security,[60] a nuclear ethics of common security orders is grounded on an inclusive duty to find "security with" one's adversaries. Thus, while the motive of national defense arises from the just demand for protection against aggression in a hostile world, the motive of common security arises from a broader and perhaps deeper concern for the survival and security of all whose fate is threatened by nuclear accident or war. Regarding justice as a motive of national defense, it is important to recall Walzer's contention:

> Nuclear weapons explode the theory of just war. . . . Nuclear war is and will remain morally unacceptable, and there is no case for its rehabilitation. Because it is unacceptable, we must seek out ways to prevent it, and because deterrence is a bad way, we must seek out others.[61]

Walzer had previously emphasized the moral necessity of national self-defense in supreme emergency conditions. Nonetheless, he correctly criticized the justifications of obliteration bombing in World War II and the use of the atomic bomb against Hiroshima as morally vacuous.[62] Accordingly, he rightly contended that it is impossible to ultimately defend the use of nuclear weapons on the principles of just war theory. Indeed, the only conclusion that just war theory and morality in general can deliver on the use of nuclear weapons is absolute condemnation. And, as chapter 4 argued, any version of political liberalism that is committed to the priority of individual human and civil rights cannot condone indiscriminate and disproportionate uses of force. Therefore, it is morally imperative to replace nuclear deterrence with an alternative means of preventing nuclear or conventional aggression because it is wrong to threaten an action whose very use introduces moral nihilism. There must be such an alternative means, one that morality and politics (or, recalling Rousseau, right and utility) can equally endorse.

A moral skeptic might object that, even if the introduction of nuclear weapons is impermissible on just war theoretic or on moral grounds, there are still adequate prudential reasons for their use if supreme emergencies arise. Prior to an actual defensive use of nuclear weapons, it is unwise to rule out the possibility that it might restore deterrence and strategic stability. In response to this objection, it is important to return to Kant's sixth preliminary article of perpetual peace: "No state at war with another shall allow itself such acts of hostility as would have to make mutual trust impossible during a future peace."[63] It is reasonable to interpret this article as prudential insofar as it proscribes unwise courses of action that would result in a deficit of the kind of mutual trust necessary to forge a future durable peace. Moreover, the article's implicit deontological content regarding the right of the adversary to exist does not invalidate or weaken the article's prudential content. In the end, Kant's subsequent argument for this article anticipates Walzerian considerations of the supreme emergency and the immorality of obliteration warfare (i.e., nuclear use) while casting doubt on the coherence of *jus ad bellum* claims by any state in the anarchy condition (just as Nardin's interpretation of Kant as a theorist of justice confirms).

It is useful to dwell a bit longer on this point by examining Kant's remarks on the sixth preliminary article and then reconstructing it in the service of the maxim under current consideration. Kant argues,

> For some trust in the enemy's way of thinking must still remain even in the midst of war, since otherwise no peace could be concluded and the hostilities would turn into a war of extermination; war is, after all, only the regrettable expedient for asserting one's right by force in a state of nature (where there is no court that could judge with rightful force); in it neither of the two parties can be declared an unjust enemy (since that already presupposes a judicial decision), but instead the outcome of the war (as in a so-called judgment of God) decides on whose side the right is; but a punitive war between states is not thinkable (since there is no relation of a superior to an inferior between them). From this it follows that a war of extermination, in which the simultaneous annihilation of both parties and with it of all right as well can occur, would let perpetual peace come about only in the vast graveyard of the human race. Hence a war of this kind, and so too the use of means that lead to it, must be absolutely forbidden.[64]

Kant's remarks highlight a contrast between two scenarios. One is where adversaries seek peace by means of mutual trust, and the other is where adversaries are committed to the total defeat of their enemy and a corresponding peace-by-domination. The latter element of the argument might be reconstructed thusly:

1. The conditions of international anarchy do not make it possible to declare authoritatively that any of the warring states lacks just cause. Thus,
2. Regrettably, any state in an anarchic world order has a right to go to war to preserve its rights that cannot otherwise be adjudicated. However,

3. It is impossible to imagine that any state going to war in these conditions can rightfully claim to be punishing an enemy for their unjust actions. Sadly,

4. If both warring parties believe they must punish their enemy to the greatest extent possible, then their actions will lead to a nihilistic war of extermination and, possibly, to the end of humanity. Therefore,

5. A war of extermination and the use of extermination weapons must be absolutely forbidden.

If this partial reconstruction satisfactorily captures Kant's thinking, then it follows that Walzer and Kant's views converge on the immorality of wars of extermination and the means of fighting them. Therefore, it is reasonable to maintain that "right and utility" or "justice and prudence" concur that nuclear use is impermissible under any circumstances, even as a response to nuclear aggression.

Maxim 3: Each state should stigmatize nuclear weapons.

This third maxim is a necessary normative component for the success of the second maxim requiring the nonuse of nuclear weapons in all circumstances. It reiterates in part the moral logic expressed by some just war and moral consequentialist commentators that nuclear restraint depends significantly upon refusing the temptation to depict nuclear weapons as just more powerful conventional weapons.[65] As Walzer had stated, "We threaten evil [i.e., nuclear reprisal] in order not to do it, and the doing of it would be so terrible that the threat seems in comparison to be morally defensible."[66] The practice that Walzer stigmatizes here is nuclear aggression, but the practice of threatening nuclear defense has not been stigmatized given the perceived need of nuclear threats for national security. For NWS adversaries that have not yet formed common security relationships, Walzer's stigmatization of nuclear aggression is as far as they can go. However, this third maxim seeks to stigmatize the element of the just war and consequentialist logic that has been acceptable so far: that is, that nuclear aggression must be met by nuclear responses. To do that, the belief that nuclear aggression must be met with nuclear reprisal must also become taboo.[67] This is best achieved if nuclear weapons themselves are stigmatized.

In this vein, the third maxim reiterates the Kantian-inspired argument that nuclear weapons are clearly beyond the pale when it comes to preserving human rights. Nuclear use is utterly inconsistent with the preservation of human rights of individual lives of innocent victims, and the threat of nuclear use constitutes innocent people as nuclear hostages against their consent.[68] Thus, the nuclear taboo must extend beyond the current prohibition on nuclear use to the prohibition on nuclear weapons possession *per se*.

It is thus important to recall the relationship between the moral practice of stigmatization and the success of an international legal prohibition regime. According to Ethan Nadelmann, an exploitative or harmful practice that nonetheless benefits its practitioners (e.g., chattel slavery, prostitution, piracy) is taken as normal and acceptable until the quantity or quality of that suffering is sufficient to produce significant opposition.[69] The process of transforming this fledgling but significant opposition into an international prohibition regime must involve stigmatization at an early stage. Unless an exploitative or harmful practice (and the objects used for that practice) is effectively and broadly stigmatized on moral grounds, efforts to criminalize it and punish practitioners will not succeed. The HINW is an instrument of moral stigmatization of nuclear weapons, as is the TPNW, but, as Tom Sauer relates, the N5 have so far prevented the success of the stigmatization efforts.[70]

The N5's success at countering stigmatization remains resilient precisely because it rests on the deeply entrenched "security against" assumptions that continue to drive international politics. Thus, it is once necessary to emphasize that the third maxim has its greatest moral and political force in relation to the first maxim of common security. Once nuclear-armed adversaries are committed to their mutual security, then the effective stigmatization of nuclear weapons can succeed.

Maxim 4: Each measure taken by states to address the questions of nuclear arms control and disarmament must conform to principles of fairness.

Mindful of Hardin's criticism that social contractarianism has failed to adequately motivate some actors to contribute their fair share to the production of common social goods, this maxim prescribes that the principles of fairness must become a central part of nuclear arms control and disarmament regime efforts. Regimes in which principles of fairness are embedded are much more likely to endure beyond the lifetimes of the officials responsible for their coming into force. Security cooperation is difficult to obtain; it is even more difficult to sustain beyond two or three decades if counter-pressures arise.[71] Thus, it cannot be morally or politically responsible to conclude a nuclear arms control or disarmament process only to discover later that its lifespan was limited to only one or two generations. Rather, nuclear regimes must be structured such that the justice and fairness of the present arrangements endure as they encounter the inevitable evolution of political conditions over long periods of time. This is to say, nuclear arms control and disarmament regimes must be concluded and maintained in ways that we might plausibly anticipate future generations would find just and fair.

Applying this maxim to the current international nuclear order would necessitate the elimination of the central inequality in the NPT regime—the difference between nuclear have and nuclear have-not states. One remedy for this inequality is that the N5 adhere to their Article VI commitments as defined by the 13 Points and the 2010 Action Plan. To do so would communicate to all NPT states-parties that the N5 regard themselves as "soon-to-be-NNWS." Admittedly, the serious challenges relating to the nuclear-armed states of India, Pakistan, Israel, and North Korea would remain. Nevertheless, this approach would be consistent with the principles of equality and reciprocity. It is also consistent with the Bodo Ethics related to the justice of promise keeping, and it has the virtue of becoming a precedent-setting example, as efforts are made to include the non-NPT states in these regimes.

Another remedy for which I have argued elsewhere is the mass exodus of the NNWS from the NPT regime.[72] This argument rests on the social contractarian premise that an international arms control and disarmament agreement, such as the NPT, is a voluntary expression of the collective will of the states-parties. In the NPT's case, the aims include nuclear nonproliferation, nuclear disarmament, and the continuation of peaceful nuclear energy. Their argument's corresponding moral premise is that each party's voluntary assent to the treaty regime constitutes a binding obligation to honor their obligations under the general norm of *pacta sunt servanda*. On the basis of these two premises, the argument contends that N5 have consistently violated their Article VI disarmament obligations as defined by the 13 Points and 2010 Action Plan, and yet they rigorously enforce the nonproliferation and peaceful nuclear energy obligations on the NNWS. The institutional effect is to shift the NPT from its disarmament mission, over the objections of the majority of NNWS, and limit it to the nonproliferation and peaceful nuclear energy missions. Such a shift counts as treaty subversion. Unfortunately, the NNWS lack the military or institutional power within the NPT or United Nations to enforce the disarmament obligations on the N5. Their choice, as this previous argument has put it, is to remain in the NPT and become complicit in regime subversion to their own long- term hurt *or* to withdraw *en masse* from the NPT and avoid further complicity in regime subversion. Additionally, since the N5 care deeply about maintaining the NPT for nonproliferation purposes, it is reasonable to claim that mass withdrawal offers the NNWS political leverage in their effort to incentivize the N5 to adhere to their disarmament obligations.

Let us now cast this argument in terms of the four maxims of the nuclear ethics of common security orders. First, the N5 have maintained steadfast "security against" postures, promoted the normalization of nuclear weapons and their modernization, and retained nuclear defense and deterrence policies that embrace first-use doctrines. Each of these stances is diametrically opposed to their NPT Article VI commitments. As a matter of rudimentary ethical norms, it is fundamentally unfair for any NNWS to be compelled to adhere to their nonproliferation obligations by threat of economic or military sanction while the N5 refuse to adhere to their nuclear disarmament obligations and promises. Consequently, the principle of equality requires that each NPT state-party adhere to regime obligations and keep its promises *or*, in the absence of maxim 1 stated earlier, act without regard to regime obligations and promises by putting national interest first. Similarly, the principle of reciprocity requires that each NPT state-party honor its regime commitments once others have done so *or* refuse continued compliance if others have stopped compliance and if there is no evidence that future compliance by others will occur. Therefore, the principles of equality and reciprocity not only permit the mass exodus of NNWS from the NPT, but on a strict reading they require it.

FINAL REMARKS

This chapter advanced a nuclear ethics of common security orders as a suitable replacement for what has become the *de facto* nuclear ethical code of the *status quo*, which maintains "security against" postures permitting or requiring an interminable reliance by a handful of states on nuclear defense and deterrence. Rather than default to the disreputable Thrasymachian stance that justice is nothing more than the advantage of the stronger,[73] the nuclear ethics of common security orders prescribes the just ordering of state relations such that adversaries and allies alike find their security and survival "with" instead of "against" each other. The nuclear ethics of the nuclear *status quo* has largely privileged order over justice on the assumption that it best provides for survival and security while the pursuit of justice provides for instability and insecurity. The privileging of order *as such* over justice permits the continuation of great power security competition, where nuclear war, defense, and deterrence strategies preserve the interests and advantages of the strongest states. As chapters 2–5 have suggested, the preservation of NWS interests and advantages increasingly puts the NNWS and humanity itself at an increased risk for their survival.

It is unfortunate that the nuclear ethics of human security, especially in its latest iteration via the Humanitarian Imperative and the TPNW, has not persuaded NWS leaders and citizenries. It is fair to say that many Kantian-inspired efforts at motivating the reconsideration of nuclear defense and deterrence policies during and after the Cold War were logically forceful but also not persuasive. Thus, Hardin's criticism of the failure of social contractarianism to motivate self-interested (state) actors also applies to the cosmopolitan humanitarianism in the nuclear abolitionist movement. One corresponding aim of this book is to address this motivational problem by linking the justice of self-survival and security with the justice of other-survival and security in the construction of a common security order. Given that people's sense of the justice of their own preservation is felt more strongly than the justice of mutual security, the challenge is to cultivate in each people a much deeper and perhaps emotional appreciation for how the conditions of the nuclear age make self- and other-survival and security mutually dependent. This appreciation is a necessary condition for the construction of a common security arrangement among (nuclear) adversaries.

If that common bond among nuclear adversaries can be constructed and maintained, then it will become possible for states acting consistent with the nuclear ethics of common security orders to eventually realize a universal and irreversible nuclear abolition. The speed of nuclear abolition's realization would be dependent upon the specific courses of action that the NWS and their allies would take. Most importantly, it would depend on the speed with which durable trust relationships among such states would be formed and maintained. The moral possibilities of a nuclear-weapon-free world must not be obscured from sight, however remote a world of common security orders might appear today.

NOTES

1. See, for example, Donaldson (1985); Lee (1985).
2. See, for example, Konoe and Maurer (2014); Tutu (2014).
3. Walzer (2015 (1977), 21–33, 51–73).
4. Walzer (2015 (1977), 250–67).
5. Payne and Payne (1987); Quinlan (2009); Ramsey (1962).
6. Hobbes (1994, 78–80).
7. See, for example, Gauthier (1984); Kavka (1978).
8. Rawls (1999, 99).
9. See, for example, Ford (2018).
10. Walker (2012, xii).
11. Walker (2012, 2). Emphasis in the original.
12. Hardin (2013, 410).
13. Hardin (2013, 411–12).
14. Hardin (2013, 412).
15. For a succinct and critical review, see Norris (1998). See also Hoffe (2006); Nardin (2017); Rawls (1999 (1971)); Rawls (1996); and Rawls (1999).
16. See Booth and Wheeler (2008, chapters 4–6).
17. Kierulf (2017, 40).
18. Walker (2012).
19. Hardin (2013, 413–14).
20. Hardin (2013, 415).
21. Hardin (2013, 420).
22. Hardin (2013, 420).
23. Nardin (2017).

24. Nardin (2017, 364).
25. Nardin (2017, 359).
26. Nardin (2017, 360).
27. Nardin (2017, 362–63).
28. Kant (1795, 1996, 8:356); Nardin (2017, 362–63).
29. Nardin (2017, 367).
30. Nardin (2017, 367).
31. Nardin (2017, 368).
32. Nardin (2017, 369).
33. Druckman and Albin (2011).
34. Druckman and Albin (2011, 1140).
35. Druckman and Albin (2011, 1143–44).
36. Druckman and Albin (2011).
37. Druckman and Albin (2011, 1140–41).
38. Druckman and Albin (2011, 1157–64.).
39. Ryan (1996, 216–25).
40. Rousseau (1987a, b).
41. Rousseau (1987b, 141).
42. Hardin (2013, 412).
43. Kant (1795, 1996, 8:358).
44. Kant (1795, 1996, 8:361–63).
45. Kant (1795, 1996, 8:344–49).
46. Kant (1795, 1996, 8:349–60).
47. Kant (1795, 1996, 8:344).
48. Kant (1795, 1996, 8:346).
49. Kant (1795, 1996, 8:347).
50. On the problem of private information in rationalist explanations of war, see, for example, Fearon (1995).
51. For instance, see the discussion above on Kant and domination, of which aggression is one kind of domination. See also Walzer (2015 (1977), 1–33).
52. Rawls (1999 (1971), 96).
53. The following discussion draws on Doyle II (2015b, 41–48).
54. Booth and Wheeler (2008, 145–58).
55. Wheeler (2018).
56. Nye (1986, 99).
57. Nye (1986, 99).
58. Nye (1986, 99).
59. See, for example, Doyle II (2015d).
60. See, for example, Gauthier (1984); Quinlan (1984 (2009)); Quinlan (2009); Payne and Payne (1987).
61. Walzer (2015 (1977), 281–82).
62. Walzer (2015 (1977), 250–67).
63. Kant (1795, 1996, 8:347).
64. Kant (1795, 1996, 8:347).
65. Hehir (1975, 1986); National Conference of Catholic Bishops (1983); Nye (1986, 99).

66. Walzer (2015 (1977), 274).

67. On the general topic of the nuclear taboo, see Tannenwald (2007).

68. Lee (1985); Shue (2004).

69. Nadelmann (1990).

70. Sauer and Reveraert (2018); Sauer (2018).

71. Examples of this point—the Concert of Vienna (1815–1854) and the U.S.–Soviet Détente (*c*. 1970–1980). See Booth and Wheeler (2008, 107–23).

72. Doyle II (2009, 2017).

73. Plato (1992, 338c).

Bibliography

PRIMARY SOURCES

1968. "Treaty on the Non-Proliferation of Nuclear Weapons (NPT)." *United Nations Office for Disarmament Affairs*. Accessed April 26, 2019. https://www.un.org/disarmament/wmd/nuclear/npt/text.

2000. "2000 Review Conference of the Parties to the Treaty on the Non-Proliferation of Nuclear Weapons: Final Document." New York. Accessed October 31, 2014. http://www.un.org/disarmament/WMD/Nuclear/pdf/finaldocs/2000%20-%20NY%20-%20NPT%20Review%20Conference%20-%20Final%20Document%20Parts%20I%20and%20II.pdf.

2010. "2010 Review Conference of the Parties to the Treaty on the Non-Proliferation of Nuclear Weapons," "Final Document, Volume 1, Part 1." New York. Accessed October 31, 2014. https://undocs.org/NPT/CONF.2010/50%20(VOL.I).

Algeria, Austria, Brazil, Chile, Costa Rica, Egypt, Guyana, et al. 2018. "Humanitarian Consequences of Nuclear Weapons." Working Paper, Preparatory Committee for the 2020 Review Conference of the Parties to the Treaty on the Non-Proliferation of Nuclear Weapons. Accessed December 27, 2018. http://undocs.org/NPT/CONF.2020/PC.II/WP.9.

Austria. 2018. "Nuclear Weapons and Security: A Humanitarian Perspective." Working Paper, Preparatory Committee for the 2020 Review Conference of the Parties to the Treaty on the Non-Proliferation of Nuclear Weapons. Accessed December 27, 2018. http://undocs.org/NPT/CONF.2020/PC.II/WP.10.

Brazil. 2018. "II PrepCom to the 2020 NPT RevCon—General Debate." April 24. Accessed February 22, 2019. http://statements.unmeetings.org/media2/18559216/brazil-printer_20180424_101949.pdf.

China. 2018. "Nuclear Issues." Working Paper, Preparatory Committee for the 2020 Review Conference of the Parties to the Treaty on the Non-Proliferation of Nuclear Weapons. Accessed December 29, 2018. http://undocs.org/NPT/CONF.2020/PC.II/WP.32.

Chinese Delegation. 2017. "Statement by the Chinese Delegation at the First Session of the Preparatory Committee for the 2020 NPT Review Conference on Security Assurance." Accessed December 28, 2018. http://statements.unmeetings.org/media2/14684682/china-new-new-english-statement-on-security-assurance.pdf.

Comprehensive Nuclear-Test-Ban Treaty Organization Preparatory Commission. 2018. *Nuclear Testing.* Accessed January 7, 2019. https://www.ctbto.org/nuclear-testing/history-of-nuclear-testing/world-overview/.

Council of Delegates of the International Red Cross and Red Crescent Movement. 2011. "Council of Delegates 2011: Resolution 1–26 November 2011." Geneva. Accessed December 21, 2018. www.icrc.org/eng/resources/documents/resolution/council-delegates-resolution-1-211.htm.

Dehghani, H. E. Mr. G. Hossein. 2017. "On Nuclear Disarmament." Statement, Preparatory Committee for the 2020 Review Conference of the Parties to the Treaty on the Non-Proliferation of Nuclear Weapons. Accessed December 28, 2018. http://statements.unmeetings.org/media2/14684608/iran-cl1.pdf.

Egypt. 2018. "The Role of Nuclear Weapons in Security and Defense Doctrines." Working Paper, Preparatory Committee for the 2020 Review Conference of the Parties to the Treaty on the Non-Proliferation of Nuclear Weapons. Accessed December 28, 2018. http://undocs.org/NPT/CONF.2020/PC.II/WP.2.

European Union. 2018. "Nuclear Disarmament Verification." Working Paper, Preparatory Committee for the 2020 Review Conference of the Parties to the Treaty on the Non-Proliferation of Nuclear Weapons. Accessed December 31, 2018. http://undocs.org/NPT/CONF.2020/PC.II/WP.6.

Group of Non-Aligned States Parties to the Treaty on the Non-Proliferation of Nuclear Weapons. 2017. "Elements for a Plan of Action for the Elimination of Nuclear Weapons." Working Paper, Preparatory Committee for the 2020 Review Conference of the Parties to the Treaty on the Non-Proliferation of Nuclear Weapons. Accessed December 28, 2018. http://undocs.org/NPT/CONF.2020/PC.I/WP.23.

Haspels, Andre. 2017. "General Statement: Non-Proliferation Treaty—Preparatory Committee 2017." Statement, Vice Foreign Minister of the Kingdom of the Netherlands. Accessed December 28, 2018. http://statements.unmeetings.org/media2/14684435/netherlands-general-statement-2-may-2017-new.pdf.

International Court of Justice. 1999. "Legality of the Threat or Use of Nuclear Weapons Advisory Opinion." In *International Law, the International Court of Justice, and Nuclear Weapons*, edited by Laurence Boisson de Chazournes and Philippe Sands, 520–60. Cambridge: Cambridge University Press.

Ireland. 2018. "Impact and Environment: The Role of Gender in the Non-Proliferation Treaty." Working Paper, Preparatory Committee for the 2020 Review Conference of the Parties to the Treaty on the Non-Proliferation of Nuclear Weapons. Accessed December 28, 2018. http://undocs.org/NPT/CONF.2020/PC.II/WP.38.

National Conference of U.S. Catholic Bishops. 1983. "The Challenge of Peace: God's Promise and Our Response." A Pastoral Letter, United States Catholic Conference, Washington, D.C. Accessed April 15, 2018. http://www.usccb.org/upload/challenge-peace-gods-promise-our-response-1983.pdf.

New Zealand on behalf of the New Agenda Coalition (Brazil, Egypt, Ireland, Mexico, New Zealand and South Africa). 2018. "Article VI of the Treaty on the

Non-Proliferation of Nuclear Weapons: reiterating the urgency of its implementation." Working Paper submitted to the Preparatory Committee for the 2020 Review Conference of the Parties to the Treaty on the Non-Proliferation of Nuclear Weapons. Accessed December 22, 2018. http://undocs.org/NPT/CONF.2020/PC.II/WP.13.

Non-Proliferation and Disarmament Initiative (NPDI). 2018. "Nuclear Safeguards Standards under the Treaty on the Non-Proliferation of Nuclear Weapons." Working Paper, Preparatory Committee for the 2020 Review Conference of the Parties to the Treaty on the Non-Proliferation of Nuclear Weapons. Accessed December 31, 2018. http://undocs.org/NPT/CONF.2020/PC.II/WP.29.

Obama, Barack. 2009. Remarks by President Barack Obama in Prague as Delivered. Prague, April 5. Accessed August 31, 2015. https://www.whitehouse.gov/the-press-office/remarks-president-barack-obama-prague-delivered.

Office of the Secretary of Defense. 2018. "Nuclear Posture Review: February 2018." United States Department of Defense, Washington, D.C. Accessed August 29, 2018. http://www.lawfareblog.com/document-nuclear-posture-review-2018.

Palme, Olof. 1982a. "Statement by Olof Palme, Chairman of the Independent Commission on Disarmament and Security Issues to the Second Special Session of the General Assembly on Disarmament." *OlofPalme.org*. June 23. Accessed July 14, 2018. http://www.olofpalme.org/wp-content/dokument/820623_fn.pdf.

"St. Paul's Epistle to the Romans." In Holy Bible, New International Version. Biblica, Inc., 2011.

The New Agenda Coalition. 2017. "Cluster 1 Issues Statement by H. E. Ambassador Patricia O'Brien on Behalf of the New Agenda Coalition." Statement, Preparatory Committee for the 2020 Review Conference of the Parties to the Treaty on the Non-Proliferation of Nuclear Weapons. Accessed December 28, 2018. http://statements.unmeetings.org/media2/14684343/ireland-on-behalf-of-nac.pdf.

The White House. 2002. "The National Security Strategy of the United States of America." Accessed January 15, 2019. https://www.state.gov/documents/organization/63562.pdf.

The White House. 2015. "National Security Strategy." Washington, D.C. Accessed July 14, 2016. https://www.whitehouse.gov/sites/default/files/docs/2015_national_security_strategy.pdf.

Ulaynov, Mikhail I. 2017. "Statement at the First Session of the Preparatory Committee for the 2020 Review Conference of the States Parties to the Treaty on the Non-Proliferation of Nuclear Weapons (Cluster 1: Nuclear Disarmament)." Permanent Representative of the Russian Federation to the International Organizations in Vienna. Accessed December 22, 2018. http://statements.unmeetings.org/media2/14684470/russian-english-cluster-1-new.pdf.

United Nations General Assembly. 2017. "Treaty on the Prohibition of Nuclear Weapons." New York. Accessed February 24, 2018. https://treaties.un.org/doc/Treaties/2017/07/20170707%2003-42%20PM/Ch_XXVI_9.pdf.

United States Department of Defense. 2010. "Nuclear Posture Review Report." Defense Review. Accessed March 31, 2018. https://www.defense.gov/Portals/1/features/defenseReviews/NPR/2010_Nuclear_Posture_Review_Report.pdf.

United States of America. 2018. "Creating the Conditions for Nuclear Disarmament (CCND)." Working Paper, Preparatory Committee for the 2020 Review of the Parties to the Treaty on the Non-Proliferation of Nuclear Weapons. Accessed December 29, 2018. http://undocs.org/NPT/CONF.2020/PC.II/WP.30.

Wood, Ambassador Robert. 2017. "2017 NPT PrepCom Cluster One Statement on Disarmament." Statement, Preparatory Committee for the 2020 Review Conference of the Parties to the Treaty on the Non-Proliferation of Nuclear Weapons. Accessed December 28, 2018. http://statements.unmeetings.org/media2/14684391/united-states-new-new-cluster-1-statement-may-5-2017.pdf.

SECONDARY SOURCES

Acheson, Ray. 2015a. "Editorial: We Can Wait No Longer." *NPT News in Review*, April 27: 1. Accessed May 22, 2015. http://www.reachingcriticalwill.org/images/documents/Disarmament-fora/npt/NIR2015/No1.pdf.

Acheson, Ray. 2015b. "Editorial: Ya Basta! It's All about the Ban." *NPT News in Review*, May 22: 1–6. Accessed May 22, 2015. http://www.reachingcriticalwill.org/images/documents/Disarmament-fora/npt/NIR2015/No16.pdf.

Adler, Emmanuel, and Michael Barnett. 1998. *Security Communities*. New York: Cambridge University Press.

Ambrose, Stephen E., and Douglas G. Brinkley. 1997. *Rise to Globalism: American Foreign Policy since 1938*. New York: Penguin Books.

Anscombe, G. E. M. 1961. "War and Murder." In *Nuclear Weapons: A Catholic Response*, edited by Walter Stein, 44–52. London: Sheed and Ward.

Arms Control Association. 2018. *Nuclear Weapons: Who Has What at a Glance*. June. Accessed March 25, 2019. https://www.armscontrol.org/factsheets/Nuclearweaponswhohaswhat.

Arms Control Association. n.d. "Treaties and Agreements." Arms Control Association. Accessed May 7, 2019. www.armscontrol.org/treaties.

Arreguin-Toft, Ivan. 2009. "Unconventional Deterrence: How the Weak Deter the Strong." Chapter 9 in *Complex Deterrence: Strategy in the Global Age*, edited by T. V. Paul, Patrick M. Morgan, and James J. Wirtz, 204–21. Chicago, IL: University of Chicago Press.

Art, Robert J. 1980. "To What Ends Military Power?" *International Security* 4 (4): 3–35.

Austin, J. L. 1975. *How to Do Things with Words*. Cambridge, MA: Harvard University Press.

Baker, Peter. 2010. "Obama Expands Modernization of Nuclear Arsenal." *New York Times*, May 13. Accessed April 27, 2018. https://www.nytimes.com/2010/05/14/us/politics/14treaty.html.

Baker, Peter, and Sang-Hun Choe. 2017. "Trump Threatens 'Fire and Fury' against North Korea if It Endangers U.S." *New York Times*, August 8. Accessed August 8, 2017. https://www.nytimes.com/2017/08/08/world/asia/north-korea-un-sanctions-nuclear-missile-united-nations.html?hp&action=click&pgtype=Homepage&clickSource=story-heading&module=first-column-region&region=top-news&WT.nav=top-news&_r=0.

BBC News. 2003. "Saddam's Iraq: Key Events." *Chemical Warfare: 1983–1988.* Accessed July 10, 2018. http://news.bbc.co.uk/2/shared/spl/hi/middle_east/02/iraq_events/html/chemical_warfare.stm.

BBC News. 2019. *Doomsday Clock Frozen at Two Minutes to Apocalypse.* January 24. Accessed January 24, 2019. https://www.bbc.com/news/world-us-canada-46992025?ns_source=facebook&ns_campaign=bbcnews&ns_mchannel=social&ocid=socialflow_facebook.

Bellamy, Alex J. 2004. *Security Communities and Their Neighbours: Regional Fortresses or Global Integrators?* London: Palgrave MacMillan.

Bernstein, Paul I. 2014. "Post-Cold War US Nuclear Strategy." Chapter 4 in *On Limited Nuclear War in the 21st Century*, edited by Jeffrey A. Larsen and Kerry M. Kartchner, 80–100. Palo Alto, CA: Stanford University Press.

Bok, Sissela. "Kant's Arguments in Support of the Maxim 'Do What Is Right Though the World Should Perish.'" *Argumentation* 2 (1988): 7–25.

Booth, Ken, and Nicholas J. Wheeler. 2008. *The Security Dilemma: Fear, Cooperation, and Trust in World Politics.* New York: Palgrave Macmillan.

Bower, Adam. 2017. *Norms without Great Powers: International Law and Changing Social Expectations.* Oxford: Oxford University Press.

Boyle, Joseph. 1992. "Natural Law and International Ethics." In *Traditions of International Ethics*, edited by Terry Nardin and David R. Mapel, 112–35. Cambridge: Cambridge University Press.

Brandom, Robert B. 2000. *Articulating Reasons: An Introduction to Inferentialism.* Cambridge, MA: Harvard University Press.

Brands, Hal. 2016. *Making the Unipolar Moment: U.S. Foreign Policy and the Rise of the Post-Cold War Order.* Ithaca, NY: Cornell University Press.

Brock, Gillian. 2010. "Recent Work on Rawls's Law of Peoples: Critics versus Defenders." *American Philosophical Quarterly* 47 (1): 85–101.

Bromwich, Jonah Engel. 2017. "Doomsday Clock Moves Closer to Midnight, Signaling Concern among Scientists." *New York Times*, January 25. Accessed January 25, 2017. https://www.nytimes.com/2017/01/26/science/doomsday-clock-countdown-2017.html?module=WatchingPortal&region=c-column-middle-span-region&pgType=Homepage&action=click&mediaId=thumb_square&state=standard&contentPlacement=2&version=internal&contentCollection=w.

Buchanan, Allen. 2000. "Rawls's Law of Peoples: Rules for a Vanished Westphalian World." *Ethics* 110: 697–721.

Buzan, Barry. 1987. "Common Security, Non-Provocative Defence, and the Future of Western Europe." *Review of International Studies* 13 (4): 265–79. Accessed July 26, 2018. https://www.jstor.org/stable/pdf/20097115.pdf?refreqid=excelsior%3A81b15e567f6bc76238538577fcfce738.

Buzan, Barry, and Lene Hansen. 2009. *The Evolution of International Security Studies.* Cambridge: Cambridge University Press.

Buzan, Barry, Ole Waever, and Jaap de Wilde. 1998. *Security: A New Framework for Analysis.* Boulder, CO: Lynne Rienner.

Callendar, Harold. 1955. "Faure Sees Paris as a Nuclear Power." *New York Times*, March 17, p. 4.

Campbell, David. 1992. *Writing Security: United States Foreign Policy and the Politics of Identity*. Manchester: Manchester University Press.

Campbell, Kurt M., Robert J. Einhorn, and Mitchell B. Reiss. 2004. *The Nuclear Tipping Point: Why States Reconsider Their Nuclear Choices*. Washington, D.C.: Brookings Institution Press.

Comprehensive Nuclear-Test-Ban Treaty Organization Preparatory Commission. 2018. *Nuclear Testing*. Accessed January 7, 2019. https://www.ctbto.org/nuclear-testing/history-of-nuclear-testing/world-overview/.Coll, Alberto R. 1999. "Normative Prudence as a Tradition of Statecraft." In *Ethics and International Affairs: A Reader*, edited by Joel H. Rosenthal, 75–100. Washington, D.C.: Georgetown University Press.

Cook, Martin L. 2004. "Christian Apocalypticism and Weapons of Mass Destruction." Chapter 10 in *Ethics and Weapons of Mass Destruction: Religious and Secular Perspectives*, edited by Sohail H. Hashmi and Steven P. Lee, 200–12. Cambridge: Cambridge University Press.

Craig, Campbell, and Jan Ruzicka. 2013. "The Nonproliferation Complex." *Ethics and International Affairs* 27 (3): 329–48.

Crawford, Neta C. 2002. *Argument and Change in World Politics: Ethics, Decolonization, and Humanitarian Intervention*. Cambridge: Cambridge University Press.

Croft, Stuart. 2012. "Constructing Ontological Security: The Insecuritization of Britain's Muslims." *Contemporary Security Policy* 33 (2): 219–35.

Cronin, Bruce. 1999. *Community under Anarchy*. New York: Columbia University Press.

Cronin, Bruce. 2001. "The Paradox of Hegemony: America's Ambiguous Relations with the United Nations." *European Journal of International Relations* 7 (1): 103–130.

Davenport, Kelsey. 2018. *The Joint Comprehensive Plan of Action (JCPOA) at a Glance*. May 9. Accessed January 23, 2019. https://www.armscontrol.org/factsheets/JCPOA-at-a-glance.

Davidson, Douglas. 2010. "The Relevance and Effectiveness of the Concept of Cooperative Security in the 21st Century." *Security and Human Rights* 21 (1): 18–20. Accessed September 10, 2018. doi:10.1163/187502310791306052.

Davis, Julie Hirschfeld. 2018. "Trump Presses NATO on Military Spending, but Signs Its Criticism of Russia." *New York Times*, July 11. Accessed July 12, 2018. https://www.nytimes.com/2018/07/11/world/europe/trump-nato-summit.html?rref=collection%2Fsectioncollection%2Fpolitics&action=click&contentCollection=politics&region=stream&module=stream_unit&version=latest&contentPlacement=10&pgtype=sectionfront.

De Luce, Dan, and Robbie Gramer. 2017. "U.S. Diplomat's Resignation Signals Wider Exodus from State Department." *Foreign Policy*, December 9. Accessed September 17, 2018. https://foreignpolicy.com/2017/12/09/u-s-diplomat-resigns-warning-of-state-departments-diminished-role-diplomacy-national-security-tillerson-africa-somalia-south-sudan/.

Deudney, Daniel H. 2007. *Bounding Power: Republican Security Theory from the Polis to the Global Village*. Princeton, NJ: Princeton University Press.

Deutsch, Karl W. et al. 1957. *Political Community and the North Atlantic Area: International Organization in the Light of Historical Experience*. Princeton, NJ: Princeton University Press.

Dittmer, Joel. n.d. "Applied Ethics." *Internet Encyclopedia of Philosophy*. Accessed March 20, 2019. https://www.iep.utm.edu/ap-ethic/.

Donaldson, Thomas. 1985. "Nuclear Deterrence and Self-Defense." *Ethics* 95 (3): 537–48.

Donnelly, Jack. 1992. "Twentieth-Century Realism." Chapter 5 in *Traditions of International Ethics*, edited by Terry Nardin and David R. Mapel, 85–111. Cambridge: Cambridge University Press.

Doyle, Michael. 1986. "Liberalism and World Politics." *American Political Science Review* 80 (4): 1151–69.

Doyle, Michael W. 1983. "Kant, Liberal Legacies, and Foreign Affairs." *Philosophy and Public Affairs* 12 (3): 205–35. Accessed July 7, 2017. http://www.jstor.org/stable/2265298.

Doyle II, Thomas E. 2009. "The Moral Implications of the Subversion of the Nonproliferation Treaty." *Ethics and Global Politics* 2 (2): 131–53.

Doyle II, Thomas E. 2010a. "Reviving Nuclear Ethics: A Renewed Research Agenda for the Twenty-First Century." *Ethics and International Affairs* 24 (3): 287–308.

Doyle II, Thomas E. 2010b. "Kantian Nonideal Theory and Nuclear Proliferation." *International Theory* 2 (1): 87–112. doi:10.1017/S1752971909990248.

Doyle II, Thomas E. 2011. "Ethics, Nuclear Terrorism, and Counter-Terrorist Nuclear Reprisals: A Response to John Mark Mattox's 'Nuclear Terrorism: The Other Extreme of Irregular Warfare.'" *Journal of Military Ethics* 10: 296–308.

Doyle II, Thomas E. 2013. "Liberal Democracy and Nuclear Despotism: Two Ethical Foreign Policy Dilemmas." *Ethics and Global Politics* 6 (3): 155–74.

Doyle II, Thomas E. 2015a. "When Liberal Peoples Turn into Outlaw States: John Rawls' Law of Peoples and Liberal Nuclearism." *Journal of International Political Theory* 11 (2): 257–73.

Doyle II, Thomas E. 2015b. *The Ethics of Nuclear Weapons Dissemination: Moral Dilemmas of Aspiration, Avoidance, and Prevention*. London: Routledge.

Doyle II, Thomas E. 2015c. "Moral and Political Necessities for Nuclear Disarmament: An Applied Ethical Analysis." *Strategic Studies Quarterly* 9 (2): 19–42.

Doyle II, Thomas E. 2015d. "Hiroshima and Two Paradoxes of Japanese Nuclear Perplexity." *Critical Military Studies* 1(2): 160–77.

Doyle II, Thomas E. 2017a. "A Moral Argument for the Mass Defection of Non-Nuclear-Weapon States from the Nuclear Nonproliferation Treaty Regime." *Global Governance: A Review of Multilateralism and International Organization* 23 (1): 15–26.

Doyle II, Thomas E. 2017b. "Deontological International Ethics." In *Oxford Research Encyclopedia of International Studies*, edited by Robert A. Denemark. Oxford: Oxford University Press. Accessed April 17, 2019. doi:10.1093/acrefore/9780190846626.013.141.

Druckman, Daniel, and Cecilia Albin. 2011. "Distributive Justice and the Durability of Peace Agreements." *Review of International Studies* 37: 1137–68. doi:10.1017/SO260210510000549.

Dummett, Michael. 1984. "Nuclear Warfare." In *Objections to Nuclear Defence: Philosophers on Deterrence*, edited by Nigel Blake and Kay Pole. London: Routledge & Kegan Paul.

Dummett, Michael. 1986. "The Morality of Deterrence." *Canadian Journal of Philosophy* 12: 111–27.

Dunn, Lewis A. 2009. "The NPT: Assessing the Past, Building the Future." *Nonproliferation Review* 16 (2): 143–72.

Ellis, Anthony. 1992. "Utilitarianism and International Ethics." Chapter 8 in *Traditions of International Ethics*, edited by Terry Nardin and David R. Mapel, 158–79. Cambridge: Cambridge University Press.

Ellsberg, Daniel. 2017. *The Doomsday Machine: Confessions of a Nuclear War Planner*. New York: Bloomsbury.

Elshtain, Jean Bethke. 1985. "Reflections on War and Political Discourse: Realism, Just War, and Feminism in a Nuclear Age." *Political Theory* 13 (1) (February): 39–57.

Erlanger, Steven, and Katrin Bennhold. 2019. "Rift between Trump and Europe Is Now Open and Angry." *New York Times*, February 17. Accessed February 17, 2019. https://www.nytimes.com/2019/02/17/world/europe/trump-international-relations-munich.html?fbclid=IwAR0CunQpEo6wdwXO1btm-9qA26aHNgUi3yHz5-TWuQfDzsn_EkhuzGNKQGo.

Farrelly, Colin (ed.). 2004. *Contemporary Political Theory: A Reader*. Thousand Oaks, CA: Sage.

Fearon, James. 1995. "Rationalist Explanations for War." *International Organization* 49 (3): 379–414.

Feldman, Fred, and Brad Skow. 2015. "Desert." *The Stanford Encyclopedia of Philosophy*, Winter. Accessed January 5, 2019. http://plato.stanford.edu/archives/win2015/entries/desert/.

Fierke, K. M. 2015. *Critical Approaches to International Security*, 2nd ed. Cambridge: Polity Press.

Fisher, Max, and Katrin Bennhold. 2018. "Germany's Europe-Shaking Political Crisis over Migrants, Explained." *New York Times*, July 4. Accessed August 7, 2018. https://www.nytimes.com/2018/07/03/world/europe/germany-political-crisis.html.

Ford, Christopher A. 2018. *Arms Control and International Security: Where Next in Building a Conditions-Focused Disarmament Discourse?* U.S. Department of State Weekly Digest Bulletin. Washington, D.C.: United States Department of State. Accessed October 26, 2018. https://www.state.gov/t/isn/rls/rm/2018/286626.htm.

Freedman, Lawrence. 2004. *Deterrence*. Cambridge: Polity.

Freeman, Samuel. 2003. "Introduction: John Rawls—An Overview." In *A Cambridge Companion to Rawls*, edited by Samuel Freeman, 1–61. Cambridge: Cambridge University Press.

Fuhrmann, Matthew, and Yonatan Lupu. 2016. "Do Arms Control Treaties Work? Assessing the Effectiveness of the Nuclear Nonproliferation Treaty." *International Studies Quarterly* 60: 530–39.

Gaddis, John Lewis. 1989. *The Long Peace: Inquiries into the History of the Cold War*. Oxford: Oxford University Press.

Gady, Franz-Stefan. 2018. "Russia Inducted 80 New ICBMs in Last 5 Years." *The Diplomat*, January 4. Accessed March 25, 2019. https://thediplomat.com/2018/01/russia-inducted-80-new-icbms-in-last-5-years/.

Ganguly, Sumit, and S. Paul Kapur. 2010. *India, Pakistan, and the Bomb: Debating Nuclear Stability in South Asia*. New York: Columbia University Press.

Gauthier, David. 1984. "Deterrence, Maximization, and Rationality." *Ethics* 94 (3): 474–95.

Gavin, Francis J. 2012. *Nuclear Statecraft: History and Strategy in America's Atomic Age*. Ithaca, NY: Cornell University Press.

Gehrke, John. 2019. "John Bolton Warns North Korea Not to Test Missiles." *Washington Examiner*, March 17. Accessed April 6, 2019. https://www.washingtonexaminer.com/policy/defense-national-security/john-bolton-warns-north-korea-not-to-test-missiles.

Gitlin, Todd. 1984. "Time to Move beyond Deterrence." *The Nation*, December 22.

Gladstone, Rick. 2016. "76 Experts Urge Donald Trump to Keep Iran Deal." *New York Times*, November 14. Accessed July 11, 2018. https://www.nytimes.com/2016/11/15/world/middleeast/trump-iran-deal.html.

Golding, Bruce. 2013. "Assad Is Like Hitler: Kerry." *New York Post*, February 2. Accessed May 20, 2014. http://nypost.com/2013/09/02/assad-is-like-hitler-kerry/.

Goodin, Robert E. 1985. "Nuclear Disarmament as a Moral Certainty." *Ethics* 95 (3) (April): 641–85.

Gould, Harry D. 2013. "Rethinking Intention and Double Effect." In *The Future of Just War: New Critical Essays*, edited by Amy Eckert and Carol Gentry, 130–47. Athens: University of Georgia Press.

Gray, Colin S. 1999. *Modern Strategy*. Oxford: Oxford University Press.

Ham, Paul. 2011. *Hiroshima Nagasaki: The Real Story of the Atomic Bombings and Their Aftermath*. New York: St. Martin's Press.

Hardin, Russell. 2013. "The Priority of Social Order." *Rationality and Society* 25 (4): 407–21. doi:10.1177/1043463113496783.

Harding, Harry. 1994. "Prospects for Cooperative Security Arrangements in the Asia-Pacific Region." *Journal of Northeast Asian Studies* 13: 31–41. Accessed September 10, 2018. doi:10.1007/BF03023283.

Hashmi, Sohail H., and Steven P. Lee (ed.). 2004. *Ethics and Weapons of Mass Destruction: Religious and Secular Perspectives*. New York: Cambridge University Press.

Hass, Ryan. 2018. "Trump Did Not Solve the North Korea Problem in Singapore—In Fact, the Threat Has Only Grown." *NBC News*. April 12. Accessed January 16, 2019. https://www.nbcnews.com/think/opinion/trump-did-not-solve-north-korea-problem-singapore-fact-threat-ncna899766.

Hehir, J. Bryan. 1975. "The New Nuclear Debate: Political and Ethical Considerations." Paper presented at the Selected Papers from the Annual Meeting (American Society of Christian Ethics).

Hehir, J. Bryan. 1986. "Morality and Deterrence: A Catholic View." *Arms Control Today* 16 (4) (May/June): 19–23.

Heinze, Eric A. 2016. *Global Violence: Ethical and Political Issues*. New York: Routledge.

Herszenhorn, David M., and Lili Bayer. 2018. "Trump's Whiplash NATO Summit: President Says US Can Go It Alone If Allies Don't Meet Spending Targets." *Politico*, July 12. Accessed July 12, 2018. https://www.politico.eu/article/trump-threatens-to-pull-out-of-nato/.

Heuser, Beatrice. 1998. *Nuclear Mentalities? Strategies and Beliefs in Britain, France, and the FRG*. New York: Macmillan.

Hobbes, Thomas. 1994. *Leviathan*. Edited by Edwin Curley. Indianapolis, IN: Hackett.

Hoffe, Ottfried. 2006. *Kant's Cosmopolitan Theory of Law and Peace*. Translated by Alexandra Newton. New York: Cambridge University Press.

Hoffman, David E. 2009. *The Dead Hand: The Untold Story of the Cold War Arms Race and Its Dangerous Legacy*. New York: Anchor Books.

Hoffmann, Stanley. 2017. "The Uses and Limits of International Law." In *International Politics: Enduring Concepts and Contemporary Issues*, edited by Robert J. Art and Robert Jervis, 176–81. New York: Pearson.

Holmes, Marcus. 2018. *Face-to-Face Diplomacy: Social Neuroscience and International Relations*. Cambridge: Cambridge University Press.

Holst, Karen. 2011. "'Palme's Political Legacy 'Put Sweden on the Map.'" *The Local*, February 28. Accessed July 16, 2018. https://www.thelocal.se/20110228/32314.

Hooker, Brad. 2016. "Rule Consequentialism." *Stanford Encyclopedia of Philosophy*. Winter, edited by Edward N. Zalta. Accessed January 9, 2019. https://plato.stanford.edu/archives/win2016/entries/consequentialism-rule/.

Huntley, Wade L. 2010. "The Abolition Aspiration." *Nonproliferation Review* 17 (1): 139–59.

Intondi, Vincent J. 2015. *African Americans against the Bomb: Nuclear Weapons, Colonialism, and the Black Freedom Movement*. Stanford, CA: Stanford University Press.

Ion, Dora. 2012. *Kant and International Relations Theory: Cosmopolitan Community-Building*. London: Routledge.

Jentleson, Bruce W. 2010. *American Foreign Policy*, 4th ed. New York: W. W. Norton & Co.

Johnson, Rebecca. 2005. "Politics and Protection: Why the 2005 NPT Review Conference Failed." *Acronym Institute for Disarmament Diplomacy*. November 1. Accessed December 28, 2018. http://www.acronym.org.uk/old/dd/dd80/80npt.htm.

Jones, Christopher M., and Kevin P. Marsh. 2014. "The Odyssey of the Comprehensive Nuclear-Test-Ban Treaty." *Nonproliferation Review* 21 (2): 207–27.

Jones, Frank L. 2013. "The High Priest of Deterrence: Sir Michael Quinlan, Nuclear Weapons, and the Just War Tradition." *Logos* 16 (3) (Summer): 15–43.

Kahn, Herman. 1985. *Thinking about the Unthinkable in the 1980s*. New York: Touchstone.

Kant, Immanuel. 1996a. "An Answer to the Question: What Is Enlightenment?" In *The Cambridge Edition of the Works of Immanuel Kant: Practical Philosophy*, edited and translated by Mary J. Gregor, 11–22. Cambridge: Cambridge University Press.

Kant, Immanuel. 1996b. "Groundwork of the Metaphysics of Morals." In *The Cambridge Edition of the Works of Immanuel Kant: Practical Philosophy*, edited and translated by Mary J. Gregor, 37–108. Cambridge: Cambridge University Press.

Kant, Immanuel. 1996c. "The Metaphysics of Morals." In *The Cambridge Edition of the Works of Immanuel Kant: Practical Philosophy*, edited and translated by Mary J. Gregor, 353–604. Cambridge: Cambridge University Press.

Kant, Immanuel. 1996d. "Toward Perpetual Peace." In *The Cambridge Edition of the Works of Immanuel Kant: Practical Philosophy*, translated by Mary J. Gregor, edited by Mary J. Gregor and Allen Wood. Cambridge: Cambridge University Press.

Kavka, Gregory. 1978. "Some Paradoxes of Deterrence." *Journal of Philosophy* 75 (6) (June): 285–302.

Kierulf, John. 2017. *Disarmament under International Law*. Montreal & Kingston: McGill-Queen's University Press.

Konoe, Tadateru, and Peter Maurer. 2014. "Remembering Hiroshima: Nuclear Disarmament Is a Humanitarian Imperative." International Committees of the Red Cross and Red Crescent, Geneva. Accessed October 2, 2014. https://www.icrc.org/eng/resources/documents/statement/2014/08-06-japan-hiroshima-atomic-bomb.htm.

Korsgaard, Christine M. 1996. *Creating the Kingdom of Ends*. Cambridge: Cambridge University Press.

Kraut, Richard. 1992. "The Defense of Justice in Plato's Republic." Chapter 10 in *The Cambridge Companion to Plato*, edited by Richard Kraut, 311–37. Cambridge: Cambridge University Press.

Krauthammer, Charles. 1990/1991. "The Unipolar Moment." *Foreign Affairs* 70 (1): 23–33.

Kuhn, Ulrich. 2018. "Deterrence and Its Discontents." *Bulletin of the Atomic Scientists* 74 (4): 248–54. Accessed July 4, 2018. doi:10-1080/009634402.2018.1486613.

Kulacki, Gregory. 2018. "China's Nuclear Force: Modernizing from Behind (2018)." *Union of Concerned Scientists*. January. Accessed March 25, 2019. https://www.ucsusa.org/nuclear-weapons/us-china-relations/nuclear-modernization.

Kydd, Andrew. 2005. *Trust and Mistrust in International Relations*. Princeton, NJ: Princeton University Press.

Kydd, Andrew H. 2000. "Trust, Reassurance, and Cooperation." *International Organization* 54 (2): 325–57.

Lackey, Douglas. 1984. *Moral Principles and Nuclear Weapons*. Totowa, NJ: Rowman & Allanheld.

Lake, David A. 1992. "Powerful Pacifists: Democratic States and War." *American Political Science Review* 86 (1): 24–37.

Landler, Mark. 2018. "Trump Abandons Iran Nuclear Deal He Long Scorned." *New York Times*, May 8. Accessed January 16, 2019. https://www.nytimes.com/2018/05/08/world/middleeast/trump-iran-nuclear-deal.html.

Lankov, Andrei. 2016. "The Reason North Korea Developed Nuclear Weapons: Survival." *The National Interest*, October 18. Accessed April 16, 2019. https://nationalinterest.org/blog/the-buzz/the-reason-north-korea-developed-nuclear-weapons-survival-18095.

Larsen, Jeffrey A., and Kerry M. Kartchner. 2014. *On Limited Nuclear War in the 21st Century*. Palo Alto, CA: Stanford University Press.

Larson, Deborah W. 1997. *Anatomy of Mistrust: U.S.-Soviet Relations during the Cold War*. Ithaca, NY: Cornell University Press.

Lee, Stephen P. 1985. "The Morality of Nuclear Deterrence: Hostage Holding and Consequences." *Ethics* 95 (4): 549–66.

Lee, Steven P. 1993. *Morality, Prudence, and Nuclear Weapons*. Cambridge: Cambridge University Press.

Lennane, Richard. 2014. "Ban the Bomb? An Australian Response." *Bulletin of the Atomic Scientists*. Accessed June 5, 2016. doi:10.1177/0096340214555079

Lorber, Eric B. 2016. "President Trump and the Iran Nuclear Deal." *Foreign Policy*, November 16. Accessed February 13, 2017. http://foreignpolicy.com/2016/11/16/president-trump-and-the-iran-nuclear-deal/.

Lupovici, Amir. 2012. "Ontological Dissonance, Clashing Identities, and Israel's Unilateral Steps towards the Palestinians." *Review of International Studies* 38 (4): 809–33. Accessed February 27, 2019. https://www.jstor.org/stable/41681491.

Lyttle, Bradford. 1983. *The Flaw in Deterrence*. Chicago, IL: Midwest Pacifist Publishing Center.

MacFhionnbhairr, Darach. 2004. "The New Agenda Coalition." Chapter 16 in *Nuclear Disarmament in the Twenty-First Century*, edited by Wade L. Huntley, Kazumi Mizumoto, and Mitsuru Kurosawa, 275–88. Hiroshima: Hiroshima Peace Institute.

Martin, Susan B. 2004. "Realism and Weapons of Mass Destruction: A Consequentialist Analysis." Chapter 4 in *Ethics and Weapons of Mass Destruction: Religious and Secular Perspectives*, edited by Sohail H. Hashmi and Steven P. Lee, 96–110. New York: Cambridge University Press.

Maoz, Zeev, and Bruce Russett. 1993. "Normative and Structural Causes of Democratic Peace, 1946–1986." *American Political Science Review* 87 (3): 624–38.

Mearsheimer, John J. 2014 (2001). *The Tragedy of Great Power Politics*. New York: W. W. Norton.

Mitzen, Jennifer. 2006a. "Anchoring Europe's Civilizing Identity: Habits, Capabilities, and Ontological Security." *Journal of European Public Policy* 13 (2): 270–85.

Mitzen, Jennifer. 2006b. "Ontological Security in World Politics: State Identity and the Security Dilemma." *European Journal of International Relations* 12 (6): 341–70.

Molder, Holger. 2011. "The Cooperative Security Dilemma in the Baltic Sea Region." *Journal of Baltic Studies* 42 (2): 143–68. doi:10.1080/01629778.2011.569063.

Morgan, Patrick M. 2003. *Deterrence Now*. Cambridge: Cambridge University Press.

Morgan, Patrick M. 2009. "Collective-Actor Deterrence." Chapter 7 in *Complex Deterrence: Strategy in the Global Age*, edited by T. V. Paul, Patrick M. Morgan, and James J. Wirtz, 158–82. Chicago, IL: University of Chicago Press.

Morgan, Patrick M., and T. V. Paul. 2009. "Deterrence among Great Powers in an Era of Globalization." Chapter 11 in *Complex Deterrence: Strategy in the Global Age*, edited by T. V. Paul, Patrick M. Morgan, and James J. Wirtz, 277–303. Chicago, IL: University of Chicago Press.

Mukhatzhanova, Gaukar. 2014. "Rough Seas Ahead: Issues for the 2015 NPT Review Conference." *Arms Control Today*, April 1: 1–9. Accessed May 15, 2015. http://www.armscontrol.org/act/2014_04/Rough-Seas-Ahead_Issues-for-the-2015-NPT-Review-Conference.

Nadelmann, Ethan A. 1990. "Global Prohibition Regimes: The Evolution of Norms in International Society." *International Organization* 44 (4): 479–526.

Nakamitsu, Izumi. 2018. "International Women's Day 2018: The Women Who Have Shaped Policies on Nuclear Weapons." *Teen Vogue*, March 8.

Nardin, Terry. 2017. "Kant's Republican Theory of Justice and International Relations." *International Relations* 31 (3): 357–72. doi:10.1177/0047117817723064.

Nolan, Janne E. 1999. *An Elusive Consensus: Nuclear Weapons and American Security after the Cold War*. Washington, D.C.: Brookings Institution Press.

Norris, Christopher W., ed. 1998. *The Social Contract Theorists: Critical Essays on Hobbes, Locke, and Rousseau.* Lanham, MD: Rowman & Littlefield.

Norris, Robert, and Hans M. Kristensen. 2010. "Global Nuclear Weapons Inventories: 1945–2010." *Bulletin of the Atomic Scientists*, July 1. Accessed July 27, 2017. doi:10.2968/066004008.

Nuclear Threat Initiative (NTI). 2018. *Non-Proliferation and Disarmament Initiative (NPDI).* May 31. Accessed December 31, 2018. https://www.nti.org/learn/treaties-and-regimes/non-proliferation-and-disarmament-initiative-npdi/.

Nye, Joseph S. 1986. *Nuclear Ethics.* New York: The Free Press.

Ogilvie-White, Tanya. 2011. *On Nuclear Deterrence: The Correspondence of Sir Michael Quinlan.* London: Routledge.

O'Neill, Onora. 1986. "The Public Use of Reason." *Political Theory* 14 (4): 523–55.

Orend, Brian. 2013. *The Morality of War*, 2nd ed. Buffalo, NY: Broadview Press.

Palme, Olof. 1982b. *Common Security: A BluePrint for Survival.* London: Pan Books.

Paul, T. V. 2009. *The Tradition of Non-Use of Nuclear Weapons.* Stanford, CA: Stanford University Press.

Payne, Keith B. 2018. "Deterrence: The 2018 NPR, Deterrence Theory and Policy." National Institute for Public Policy Information Series, National Institute for Public Policy. Accessed September 13, 2018. http://www.nipp.org/national-institute-press/information-series/.

Payne, Keith B., and Karl I. Payne. 1987. *A Just Defense: The Use of Force, Nuclear Weapons and Our Conscience.* Portland, OR: Multnomah Press.

Pelopidas, Benoit. 2015. "A Bet Portrayed as a Certainty: Reassessing the Added Deterrent Value of Nuclear Weapons." In *The War That Must Never Be Fought: Dilemmas of Nuclear Deterrence*, 5–55. Stanford, CA: Hoover Institution Press.

Plato. 1992. *The Republic.* Translated by G. M. A. Grube. Indianapolis, IN: Hackett.

Pratt, Simon Frankel. 2017. "A Relational View of Ontological Security in International Relations: A Theory Note." *International Studies Quarterly* 61: 78–85. Accessed November 15, 2018. doi:10.1093/isq/sqw038.

Quinlan, Michael. 2009. *Thinking about Nuclear Weapons: Principles, Problems, Prospects.* London: Oxford University Press.

Quinlan, Michael. 2009 (1984). "The Strategic Uses of Nuclear Weapons." In *Thinking about Nuclear Weapons*, 184–89. Oxford: Oxford University Press.

Quinlan, Michael. 2009 (1981). "Nuclear Weapons and Preventing War." In *Thinking about Nuclear Weapons: Principles, Problems, Prospect*s, 181–88. London: Oxford University Press.

Ramsey, Paul. 1962. "The Case for Making 'Just War' Possible." In *Nuclear Weapons and the Conflict of Conscience*, edited by John C. Bennett, 143–72. New York: Charles Scribner's.

Ramsey, Paul. 2002 (1968). *The Just War: Force and Political Responsibility.* New York: Rowman & Littlefield.

Rathbun, Brian C. 2012. *Trust in International Cooperation: The Creation of International Security Institutions and the Domestic Politics of American Multilateralism.* New York: Cambridge University Press.

Rawls, John. 1996. *Political Liberalism.* New York: Columbia University Press.

Rawls, John. 1999. *The Law of Peoples.* Cambridge: Cambridge University Press.

Rawls, John. 1999 (1971). *A Theory of Justice, Revised Edition*. Cambridge, MA: Harvard University Press.

Reidy, David A. 2004. "Rawls on International Justice: A Defense." *Political Theory* 32 (3): 291–319. Accessed February 27, 2019. https://www.jstor.org/stable/4148156.

Reif, Kingston. 2018. *Trump to Withdraw from INF Treaty*. November. Accessed January 23, 2019. https://www.armscontrol.org/act/2018-11/news/trump-withdraw-us-inf-treaty.

Roberts, Brad. 2016. *The Case for U.S. Nuclear Weapons in the 21st Century*. Stanford, CA: Stanford University Press.

Rose, Frank A. 2018. "'Two Halves of the Same Walnut': The Politics of New START Extension and Strategic Nuclear Modernization." *The Brookings Institution*. August 30. Accessed January 16, 2019. https://www.brookings.edu/blog/order-from-chaos/2018/08/30/two-halves-of-the-same-walnut-the-politics-of-new-start-extension-and-strategic-nuclear-modernization/.

Rousseau, Jean-Jacques. 1987a. "Discourse on the Origin of Inequality." In *The Basic Political Writings*, edited and translated by Donald A. Cress, 25–110. Indianapolis, IN: Hackett.

Rousseau, Jean-Jacques. 1987b. "On the Social Contract." In *The Basic Political Writings*, edited and translated by Donald A. Cress, 141–227. Indianapolis, IN: Hackett.

Russett, Bruce. 1998. "A Neo-Kantian Perspective: Democracy, Interdependence and International Organizations in Building Security Communities." In *Security Communities*, edited by Emmanuel Adler and Michael Barnett, 368–94. Cambridge: Cambridge University Press.

Russett, Bruce, with William Antholis, Carol R. Ember, Melvin Ember, and Zeev Maoz. 1993. *Grasping the Democratic Peace: Principles for a Post-Cold War World*. Princeton, NJ: Princeton University Press.

Ryan, Alan. 1996. "Hobbes's Political Philosophy." Chapter 9 in *The Cambridge Companion to Hobbes*, edited by Tom Sorell, 208–45. Cambridge: Cambridge University Press.

Sagan, Scott D. 1996–1997. "Why Do States Build Nuclear Weapons? Three Models in Search of a Bomb." *International Security* 21 (3): 54–86.

Sagan, Scott D. 2000. "The Commitment Trap: Why the United States Should Not Use Nuclear Threats to Deter Biological and Chemical Weapons Attacks." *International Security* 24 (4): 85–115.

Sagan, Scott D., and Kenneth N. Waltz. 2003. *The Spread of Nuclear Weapons: A Debate Renewed*. New York: W. W. Norton.

Sanger, David E., and Rick Gladstone. 2017. "Contradicting Trump, U.N. Monitor Says Iran Complies with Nuclear Deal." *New York Times*, August 31. Accessed July 11, 2018. https://www.nytimes.com/2017/08/31/world/middleeast/un-nuclear-iran-trump.html.

Sanger, David E., and William J. Broad. 2017. "Trump Forges Ahead on Costly Nuclear Overhaul." *New York Times*, August 27. Accessed August 28, 2017. https://www.nytimes.com/2017/08/27/us/politics/trump-nuclear-overhaul.html?h

p&action=click&pgtype=Homepage&clickSource=story-heading&module=first-column-region&region=top-news&WT.nav=top-news.

Santas, Gerasimos. 2001. *Goodness and Justice: Plato, Aristotle, and the Moderns.* Malden, MA: Blackwell.

Sauer, Tom. 2005. *Nuclear Inertia: US Weapons Policy after the Cold War.* London: I. B. Tauris.

Sauer, Tom. 2018. *Whether You Like It or Not, the Nuclear Ban Treaty Is Here to Stay: A Reply to Brad Roberts.* Commentary: 29 March, European Leadership Network. Accessed March 31, 2018. https://www.europeanleadershipnetwork. org/commentary/whether-you-like-it-or-not-the-nuclear-ban-treaty-is-here-to-stay-a-reply-to-brad-roberts/.

Sauer, Tom, and Mathias Reveraert. 2018. "The Potential Stigmatizing Effect of the Treaty on the Prohibition of Nuclear Weapons." *The Nonproliferation Review*, 1–19. Accessed January 10, 2019. doi:10.1080/10736700.2018.1548097.

Sayre-McCord, Geoff. 2017. "Moral Realism." Edited by Edward N. Zalta. *The Stanford Encyclopedia of Philosophy.* Accessed April 2, 2019. https://plato.stanford. edu/archives/fall2017/entries/moral-realism/.

Scanlon, Thomas M. 1998. *What We Owe Each Other.* Cambridge, MA: Belknap Press of Harvard University Press.

Scarry, Elaine. 2014. *Thermonuclear Monarchy: Choosing between Democracy and Doom.* New York: W. W. Norton.

Schell, Jonathan. 1982. *The Fate of the Earth.* New York: Alfred A. Knopf.

Schell, Jonathan. 1984. *The Abolition.* New York: Knopf.

Schelling, Thomas C. 2008 (1966). *Arms and Influence, with a New Preface and Afterword.* New Haven, CT: Yale University Press.

Schelling, Thomas C. 1980 (1960). *The Strategy of Conflict.* Cambridge, MA: Harvard University Press.

Searle, John R. 2010. *Making the Social World: The Structure of Human Civilization.* Oxford: Oxford University Press.

Sechser, Todd S., and Matthew Fuhrmann. 2017. *Nuclear Weapons and Coercive Diplomacy.* New York: Cambridge University Press.

Shue, Henry. 2004. "Liberalism: The Impossibility of Justifying Weapons of Mass Destruction." Chapter 7. In *Ethics and Weapons of Mass Destruction: Religious and Secular Perspectives*, edited by Sohail H. Hashmi and Steven P. Lee, 139–62. Cambridge: Cambridge University Press.

Sinnott-Armstrong, Walter. 1988. *Moral Dilemmas.* Oxford: Basil Blackwell.

Sjoberg, Laura (ed.). 2010. *Gender and International Security: Feminist Perspectives.* New York: Routledge.

Sklar, Judith N. 1989. "The Liberalism of Fear." In *Liberalism and the Moral Life*, edited by Nancy L. Rosenblum, 21–37. Cambridge, MA: Harvard University Press.

Smoke, Richard. 1993. *National Security and the Nuclear Dilemma: An Introduction to the American Experience in the Cold War.* New York: McGraw-Hill.

Snyder, Scott A. 2016. "North Korea's Testing Decade." *Council on Foreign Relations.* October 7. Accessed January 16, 2019. https://www.cfr.org/expert-brief/north-koreas-testing-decade.

Solingen, Etel. 2007. *Nuclear Logics: Contrasting Paths in East Asia and the Middle East*. Princeton, NJ: Princeton University Press.

Steele, Brent J. 2008. *Ontological Security in International Relations: Self-Identity and the IR State*. London: Routledge.

Tannenwald, Nina. 2007. *The Nuclear Taboo: The United States and the Non-Use of Nuclear Weapons since 1945*. Cambridge: Cambridge University Press.

Tannenwald, Nina. 2018. "How Strong Is the Nuclear Taboo Today?" *Washington Quarterly* 41 (3): 89–109.

Thakur, Ramesh. 2015. *Nuclear Weapons and International Security: Collected Essays*. New York: Routledge.

The Century Foundation. 2018. "Trump's First Year Unsettles World Politics." January 17. Accessed July 11, 2018. https://tcf.org/content/report/trump-first-year-unsettles-world-politics/?agreed=1.

The Economist. 2018. "The EU Should Get Tough on Its Illiberal Democracies." April 19. Accessed January 16, 2019. https://www.economist.com/leaders/2018/04/19/the-eu-should-get-tough-on-its-illiberal-democracies.

Tickner, J. Ann. 1991. "A Critique of Morgenthau's Principles of Political Realism." In *Gender and International Relations*, edited by Rebecca Grant and Kathleen Newland. Bloomington: Indiana University Press.

Tutu, Desmond. 2014. "Imagine a World without Nuclear Weapons." In *Banning Nuclear Weapons: An African Perspective*, edited by Arielle Denis, 4–5. International Campaign to Abolish Nuclear Weapons.

Vuori, Juha A. 2016. "Deterring Things with Words: Deterrence as a Speech Act." *New Perspectives: Interdisciplinary Journal of Central & East European Politics & International Relations* 24 (2): 23–50.

Walker, William. 2012. *A Perpetual Menace: Nuclear Weapons and International Order*. New York: Routledge.

Walt, Stephen M. 1985. "Alliance Formation and the Balance of World Power." *International Security* 9 (4): 3–43.

Waltz, Kenneth N. 2003. "More May Be Better." Chapter 1 in *The Spread of Nuclear Weapons: A Debate Renewed*, edited by Scott D. Sagan and Kenneth N. Waltz, 3–45. New York: W. W. Norton.

Walzer, Michael. 2015 (1977). *Just and Unjust Wars: A Moral Argument with Historical Illustrations*, 5th ed. New York: Basic Books.

Wang, Zheng. 2012. *Never Forget National Humiliation: Historical Memory in Chinese Politics and Foreign Relations*. New York: Columbia University Press.

Ward, Alexander. 2018. "Trump Doesn't Have a 'Nuclear Button' on His Desk. He Could Easily Attack North Korea Anyway." *Vox*, January 9. Accessed January 11, 2018. https://www.vox.com/world/2018/1/3/16844772/trump-north-korea-button-nuclear-taunt.

Weede, Erich. 1983. "Extended Deterrence." *Journal of Conflict Resolution* 27 (2) (June): 231–54.

Weiss, Thomas G., David P. Forsythe, Roger A. Coate, and Kelly-Kate Pease. 2017. *The United Nations and Changing World Politics*, 8th ed. Boulder, CO: Westview Press.

Wenar, Leif. 2013. "John Rawls." Edited by Edward N. Zalta. *The Stanford Encyclopedia of Philosophy*. Accessed November 15, 2018. http://plato.stanford.edu/archives/win2013/entries/rawls/.

Wendt, Alexander. 1999. *Social Theory of International Politics*. Cambridge: Cambridge University Press.

Wheeler, Nicholas J. 2018. *Trusting Enemies: Interpersonal Relationships in International Conflict*. Oxford: Oxford University Press.

Wilson, John. 1963. *Thinking with Concepts*. Cambridge: Cambridge University Press.

Wittner, Lawrence S. 2009. *Confronting the Bomb: A Short History of the World Nuclear Disarmament Movement*. Stanford, CA: Stanford University Press.

Woodward, Bob. 2018. *Fear: Trump in the White House*. New York: Simon & Schuster.

Wright, Tim. 2015. "The Humanitarian Push to Prohibit Nuclear Weapons." *NPT News in Review*, April 27, 4.

Zarakol, Ayse. 2010. "Ontological (In)security and State Denial of Historical Crimes: Turkey and Japan." *International Relations* 24 (1): 3–23.

Zellner, Wolfgang. 2010. "Cooperative Security—Principle and Reality." *Security and Human Rights* 21 (1): 64–68. Accessed September 10, 2018. doi:10.1163/187502310791305981.

Index

applied ethics, 14, 165, 191; and Bodo Ethics, 163, 169, 170, 179; and consequentialism, 9–13 (and antinuclear consequentialism, 46–48 (Gitlin, Todd, 46–47; Goodin, Robert, 47–48); and pro-deterrence consequentialism, 27–37 (Gauthier, David, 28–31, 46; Kavka, Gregory, 31–32, 43, 46); and deontological ethics, 10–17, 41–45, 146, 153, 165, 174, 176 (Bok, Sisela, 24–25, 38, 82; Donaldson, Thomas, 40–41, 58, 160; Dummett, Michael, 41–45, 58; Kant, Immanuel, 10–11, 23–25, 37–45, 58, 72, 80–86, 98–104, 107–10, 135, 146, 148, 159–67, 169–70, 176–77, 180 (categorical imperative, 38–39, 82, 135, 146, 165; Formula of Humanity, 39, 41, 44, 45; Formula of Universal Law, 39, 41, 45); Lee, Steven P., 43–48, 53n111, 58, 64, 146–47, 160 (Principle of the Morality of Social Institutions (PMSI), 43–45, 146–47, 152–53); Nardin, Terry, 89n112, 165–68, 176; O'Neill, Onora, 84–85; wrongful intentions principle (WIP), 36, 105); and just war theory, 10, 29, 35, 36, 41, 100, 107, 175 (Anscombe, Elizabeth, 25, 27, 33, 36, 41; discrimination (a.k.a., principle of noncombatant immunity), 29, 33–36, 38, 40–42, 45, 55, 61, 64, 91, 105, 106, 115, 121, 146; doctrine of double effect, 35–36; Hehir, J. Bryan, 33–35, 109; jus ad bellum, 11, 13, 14, 27, 35, 39, 49, 53n111, 55, 106, 115, 146, 160, 170, 176; jus in bello, 11, 29, 33–37, 40–42, 49, 53n111, 55, 58, 61, 64, 104–11, 115, 146; Payne, Keith, 27, 29, 33, 38, 66; Quinlan, Michael (Sir), 29, 35–37, 39, 42, 44, 55, 58; Ramsey, Paul, 33–34, 38; U.S. Catholic Bishops, 23, 27, 34, 35, 55, 58, 109 (Challenge of Peace (1983), 27, 34); Walzer, Michael, 15, 23, 28, 34, 36–42, 58, 92, 105–9, 115, 139, 145–46, 175–77 (supreme emergency condition, 37, 39, 40, 58, 105, 106, 139, 146, 147, 175, 176)); and moral incoherence, 49, 86, 91, 107–12, 115; and moral justification, 13, 17, 19, 28, 32, 55, 67, 91, 98, 106, 109, 121, 147; and noncognitivism (emotivism), 8–9)

Deudney, Daniel, x, 112

Gavin, Francis, 50n4, 52n10, 119n110, 138

About the Author

Thomas E. Doyle II is Associate Professor of Political Science at Texas State University.

www.ingramcontent.com/pod-product-compliance
Lightning Source LLC
Chambersburg PA
CBHW030649270326
41929CB00007B/278